LISA STEELE

101
CHICKEN KEEPING
HACKS

FROM FRESH EGGS DAILY

Quarto.com

© 2018 Quarto Publishing Group USA Inc.
Text © 2018 Lisa Steele

Photography © 2018 Peg Keyser/CoopduJour Photography

First published in 2018 by Voyageur Press, an imprint of The Quarto Group,
100 Cummings Center, Suite 265-D, Beverly, MA 01915, USA.
T (978) 282-9590 F (978) 283-2742

Voyageur Press titles are also available at discount for retail, wholesale, promotional, and bulk
purchase. For details, contact the Special Sales Manager by email at specialsales@quarto.com or by
mail at The Quarto Group, Attn: Special Sales Manager, 100 Cummings Center, Suite 265-D, Beverly,
MA 01915, USA.

11

ISBN: 978-0-7603-6063-7

Digital edition published in 2018
eISBN: 978-0-7603-6064-4

Library of Congress Cataloging-in-Publication Data

Names: Steele, Lisa, 1965- author.
Title: 101 chickenkeeping hacks from Fresh eggs daily : tips, tricks, and
   ideas for you and your hens / by Lisa Steele.
Other titles: One hundred one chickenkeeping hacks from Fresh eggs daily
Description: Minneapolis, Minnesota : Voyageur Press, 2018. | Includes index.
Identifiers: LCCN 2018015040 | ISBN 9780760360637 (pb)
Subjects: LCSH: Chickens.
Classification: LCC SF487 .S814 2018 | DDC 636.5--dc23
LC record available at https://lccn.loc.gov/2018015040

Acquiring Editor: Thom O'Hearn
Project Manager: Jordan Wiklund
Art Director: Cindy Samargia Laun
Cover and Page Design: Amy Sly
Photography: Peg Keyser

Printed in China

TO MY GRANDFATHER,

who took me on my first dump run.
He could (and did!) fix just about anything
with a roll of duct tape and some baling twine.

Table of Contents

INTRODUCTION

My grandparents were chicken farmers who didn't have a lot of extra time or money. They were no different than most of their generation; everything they did had to be economical, practical, and efficient. They didn't throw away things that broke or rusted. Instead, they fixed what they had or repurposed it. When I was young, my grandfather regularly made trips to the town dump—and I was thrilled when I could join him! We loaded up his station wagon with all sorts of treasures: scrap metal and cans that he could turn in for a few pennies, broken radios or lawnmowers that he thought he could fix, and often a toy for me that he cleaned up and repaired when we got home. My mom also loves giving things new life. A lifelong yard-saler with New England thriftiness and ingenuity, she used thrift and DIY to make sure my brother and I had the best toys in the neighborhood! (I've told her that were she fifty years younger, she would definitely be a craft-DIY blogger!)

That love of "junking" was passed down to me. Over the last decade or so that I've been raising chickens, I've tried out plenty of high-end, top-of-the-line commercial products, but none gave me the satisfaction of pulling a MacGyver and rigging up my own creations to make my chickens' lives a little better without spending a dime. No doubt you've already learned that backyard chicken keeping, like most things in life, can be as simple or as complicated as you choose to make it. You can spend as much or as little money on your feathered flock as you wish. Take coops, for example. If you're handy, you might find yourself building a coop for

your chickens. If you're looking to save money, you can find a used shed or playhouse that can be converted into a serviceable coop. Not handy at all? Plan on spending anywhere from a few hundred to a few thousand dollars on a chicken coop, depending on the level of luxury you want to give your fluffy pets.

My goal with this book is to be practical and to stay true to the basic concept of a "hack." Hacks are all about improvisation and creativity. They solve problems in ways that save money, save time, or simply change the way something has always been done. I've heard a hack referred to as "an appropriate application of ingenuity," and that sounds about right!

In my eyes, hacks have to do more than be clever for the sake of being clever, and there's no need to invent problems. Take the chicken feeder and waterer, for example. Over the years I have tried everything from the PVC-type hookups to the treadle feeder contraptions. I've used the traditional gravity-style feeders as well as the nipple-concept waterers. However, I keep coming back to a basic enamelware roasting pan or casserole dish that you can pick up secondhand in a variety of sizes

for both feed and water. They're inexpensive and superdurable. They don't rust over time, nor does the enamelware crack in the winter. There's no need for a DIY project or hack here! You have to know a good thing when you see it and stick with the basics when it makes sense.

Much of my chicken keeping is based on old-timers' methods. I try to embrace those things that people have been doing forever in their gardens and their coops (before synthetic chemicals, before commercial products, when frugality was a way of life). Many of the hacks in this book are inspired by those wise chicken keepers.

Last but not least, I encourage you to make do when possible and repurpose what you have on hand. In many cases throughout the book, I used materials I found in our barn, garage, or scrap pile. If you don't have the exact same size, shape, color, or style that I describe, see if you can use what you do have or find something similar on the cheap. I've found that chickens rarely complain if the results are less than picture perfect.

Lisa

FEEDING YOUR CHICKENS

It's important to feed your chickens high-quality food for a healthy and balanced diet. After all, chickens expend plenty of energy on their own, then expend more energy and use additional nutrients laying their eggs. Any reputable feed company likely offers a complete line of feed, from chick starter to layer, but mixing your own feed isn't difficult—as long as you have a good recipe that offers the balance of nutrients a flock needs. If you don't want to mess with your flock's feed, there are still plenty of ways to provide your chickens nutritious treats and to save money! In this chapter, we'll explore offering everything from insects, grass, and seeds to kitchen scraps and chicken treats. You'll learn how (and why) to ferment chicken feed and how to give your chickens fresh greens year-round. You'll soon diversify your chickens' diet—and less of your food budget will go to waste as well.

HOMEMADE LAYER FEED

Chickens need a good-quality, balanced feed in order to remain in optimal health and have the energy and nutrients they need to lay those delicious fresh eggs. Tossing them some grains and seeds might work in the short term, but for a long laying life, you really need to stick to a tried-and-true mix. This recipe for layer feed comes from Diana Ambauen-Meade, founder and CEO of Scratch and Peck Feeds.

WHAT YOU'LL NEED

25 pounds peas, split or cracked

15 pounds wheat (soft white, hard red, or a combination)

15 pounds cracked corn

10 pounds oats

10 pounds barley

10 pounds fish meal

5 pounds sunflower seeds

5 pounds alfalfa meal

2 pounds sea kelp meal

1¾ pounds dicalcium phosphate

1¼ pound crushed oyster shells or limestone

YIELD: 100 POUNDS

Mix all ingredients well. It's important that the smaller additions are thoroughly integrated throughout the feed. Store in a tightly covered pail or other container to keep insects, rodents, and moisture out. Offering black soldier fly larvae to your chickens in conjunction with the ingredients in this mix will provide them methionine (an essential amino acid) and some additional calcium. Otherwise, methionine can be purchased and 0.1 pound can be added to the mix.

Flaxseed (up to 10 pounds) can also be added to the mix (or offered as a treat) to provide additional omega-3 fatty acids to your chickens—it will translate into the eggs they lay. Flaxseed should be added whole, because it loses nutrients quickly once it's ground.

Quick Hack

Buy a bag of wild birdseed that includes a mixture of seeds, nuts, grains, and even dried fruit. My chickens seem to really enjoy the woodpecker seed mixes that include peanuts, pistachios, dried cranberries, and sunflower seeds.

HOMEMADE SCRATCH MIX

If you don't want to mix your own layer feed, why not try mixing scratch grains instead? Scratch grains fall into the treat category, to be fed in moderation, and you don't need to worry about the exact nutritional content. This means you can mix and match as you wish. Typically scratch grains are fed during the cold winter months just before bedtime to provide energy and heat as your chickens' bodies work to digest the grains. Thus, they are generally higher in fat and lower in protein than feed grains.

WHAT YOU'LL NEED

Use the following list to create your own scratch-grains mix. Choose a few ingredients from each category to create a fun blend for your flock:

NUTS (SHELLED AND UNSALTED)
Peanuts, Almonds, Walnuts

SEEDS
Sunflower seeds (hulled or in shell), Safflower seeds, Sesame seeds, Flaxseed, Pumpkin seeds, Thistle seeds

GRAINS
Dried corn (whole or cracked), Rolled or steel-cut oats, Wheat, Barley, Millet, Milo

DRIED FRUIT
Raisins, Cranberries, Apricots, Apples

Mix the ingredients in a covered pail and offer some to your chickens before bed in the winter. Limit the scratch to about 10 percent of the overall diet of your flock, which would amount to about a tablespoon of scratch grains per chicken per day.

I've discovered that the fliptop shaker lid that comes on some plastic Parmesan cheese containers will also fit onto a regular-mouth Mason jar, turning it into a nice scratch dispenser! A few shakes and you'll have your chickens running for the grains as they scatter on the ground!

Quick Hack

Fill a plastic berry box (the kind with the attached lid and small holes in it) with scratch grains and let your chickens kick that around the run. It's the ultimate winter boredom buster!

TOP TWELVE
Dried Herbs for Feed

Studies have shown that eggs laid by chickens with more access to grasses and weeds are more nutritious, so it stands to reason that adding some herbs to your chickens' diet will have a similar effect. All of the culinary herbs are perfectly safe and edible for chickens, and most offer additional health benefits. If you grow herbs, think about planting a few extras for your chickens this spring! Planting and feeding fresh herbs to your penned-up chickens is a great way to add nutrition and variety to their diet while still keeping them safe from predators.

My chickens have their own herb garden right next to their coop to munch on during the growing season. In the fall before the first frost, I harvest, dry, and crumble whatever is left in the garden to add to my chickens' feed through the winter, when they don't have access to fresh grass and weeds to eat.

I have spent way more time than I should watching which herbs my chickens tend to nibble on most when given a wide variety of herbs to choose from, so I feel confident that I can pick out their favorites! These are my (or rather, my chickens') 12 favorite herbs:

» **Basil**—Basil is one of my favorite herbs to cook with and also one of my chickens' favorites! They love to wander through the herb garden and nibble on the soft, aromatic leaves during the growing season. Basil works as an antioxidant and anti-inflammatory; aids mucus-membrane, circulatory, and respiratory health; supports orange egg yolks; aids in digestive and immune-system health; and is a great source of protein, vitamin K, and iron.

» **Cilantro**—Cilantro is a funny herb in that people seem to either love it or hate it. There's something about the way the human body is wired that dictates whether we enjoy the taste of cilantro. Not so with my chickens! They always gravitate toward the cilantro patch, which I grow exclusively for them because I'm not personally a fan of this pungent herb. Cilantro is an antioxidant that builds strong bones while also being high in vitamin A for vision and vitamin K for blood clotting.

» **Dandelion**—Although my husband curses the cheerful yellow dandelions that pop up in the yard each spring, I adore seeing them. Chickens absolutely love dandelion greens, which are a general health tonic; a diuretic that improves digestive, kidney, and liver health; a laying stimulant that is high in calcium for strong eggshells; and an antioxidant and anti-inflammatory. Dandelion greens also contribute to orange egg yolks.

» **Dill**—Being of Scandinavian descent, I love cooking with dill. Unfortunately I have to fight the chickens for it! They're big fans of dill, likely for its antioxidant properties. Dill also improves respiratory and digestive health, stimulates the appetite, and promotes feather growth, so I'm happy to share a bit with my chickens!

» **Marigold**—Many commercial feed companies add marigolds to their recipes to promote vibrant orange egg yolks from the chickens who eat their feed. But marigolds are also beneficial to your flock for their antioxidant properties, as well as being a laying stimulant that supports a healthy digestive system.

» **Marjoram**—Marjoram is another nutritious herb that will enhance the color of egg yolks, act as a laying stimulant, and aid in digestive health. I grow marjoram specifically for my chickens because of how much they love to nibble on it.

» **Mint**—If you don't grow anything else, you should grow mint. It not only is pretty much indestructible and spreads like crazy, it has multiple uses in chicken keeping. I love adding it to my nesting area to deter bugs and rodents. Added to your chickens' diet, it works as an antioxidant, aids in respiratory and digestive health, improves feather growth, and is scientifically proven to result in larger eggs, thicker eggshells, and increased egg production.

» **Oregano**—Oregano has long been studied and used as a natural antibacterial. I guess that's why my chickens love to munch on it so much. Research suggests that carvacrol, which is a component of oregano oil, helps to combat coccidia, salmonella, infectious bronchitis, avian flu, blackhead, and *E. coli*; strengthens immune systems; aids in respiratory and digestive health; and acts as an anti-inflammatory. Offering oregano to your flock is so beneficial.

» **Parsley**—Out of all the culinary herbs, parsley is the one that's highest in nutritional value. Parsley is packed with vitamins A, B, and C as well as calcium and iron; aids in bone and blood-vessel development; improves circulation; works as a laying stimulant, antioxidant, and anti-inflammatory; and also aids in digestive health. If you use parsley in your own cooking, be sure to plant a bit extra for your chickens! And also feed them the stems when a recipe calls for parsley leaves.

» **Sage**—Sage is another herb that has some wonderful benefits for your chickens. Often paired with chicken in recipes, sage is thought to combat salmonella. I love frying fresh sage leaves to top a pasta dish, but fortunately my chickens are happy to just nibble on the fresh leaves. Sage is an antioxidant, antiparasitic, and antibacterial, as well as a general health promoter, immune-system booster, and laying stimulant.

» **Tarragon**—Tarragon is one of my favorite herbs, especially in scrambled eggs or an omelet. Coincidentally, my chickens love tarragon as well. Possibly because it's an antioxidant and appetite stimulant?

» **Thyme**—Thyme is a superhardy herb that is a perennial and will spread in time. This makes it the perfect herb to add to your garden for your chickens. Thyme aids in respiratory, digestive, and immune-system health, and also works as an antioxidant, antiparasitic, and laying stimulant.

Quick Hack

Don't stop with herbs! There are lots of edible flowers that your chickens will love to nibble on as well, including echinacea, bee balm, rose petals, violets, and more!

OATMEAL
A Warm Winter Treat

According to the US Department of Agriculture's Farmers' Bulletin, baby chicks fed a ration of oats will be healthier than chicks that aren't offered oats. Oats are an excellent source of antioxidants and protein, as well as vitamins and minerals, including thiamine, riboflavin, niacin, choline, calcium (which is so important—especially to laying hens), copper, iron, magnesium, and zinc. Oats have also been shown to reduce death rates in flocks, improving their general health, reducing pecking and aggression (which often lead to cannibalism in flocks), and bolstering resistance to heat exhaustion.

In addition to offering baby chicks raw oats to help with pasty butt (see page 89), I add raw rolled oats to my adult hens' daily feed. But when the temperatures drop, warm oatmeal is a nutritious treat that my chickens love. Each hen gets only a tablespoon or 2—not a lot. It's considered a treat, so it should be fed in limited amounts.

If you'd like to do the same, just heat water in a teakettle, then pour the warm water over a pan of rolled oats. I use just enough water to moisten them. By the time I get to the coop, the oatmeal has cooled enough to serve. Add a bit of cinnamon or cayenne if you'd like, or other mix-ins such as nuts, seeds, berries, dried herbs, or mealworms. Yum!

And don't stop at oats! Your chickens will also love the occasional treat of other cooked grains, including quinoa, millet, barley, or even grits!

WINTERTIME
Boredom-Busting Block

In the winter, chickens tend to get bored. They also benefit from a bit more protein and fat to help them stay warm. There are seed blocks available commercially, of course, but this homemade version is super easy to make, and your chickens will love it. Serve it to your chickens as an occasional treat that makes up less than 10 percent of their diet.

WHAT YOU'LL NEED

40-ounce plastic peanut-butter jar or similar container

¼ cup peanut butter (organic, unsalted)

¼ cup coconut oil, heated until liquified

¼ cup cold water

¼-ounce envelope unflavored gelatin

1 cup boiling water

6 cups mixture of scratch grains, seeds, cracked corn, raisins, dried fruit, mealworms, and nuts

YIELD: 1 BLOCK

With sharp scissors, cut off the top of the peanut-butter jar just below where it starts to narrow. Set the jar aside and discard the cut-off top and lid.

Whisk the peanut butter into the coconut oil until combined. Set aside.

Sprinkle the gelatin powder on top of the ¼ cup of cold water and let it sit for a minute. Pour the boiling water into a medium bowl and whisk in the gelatin mixture. Continue to whisk until the gelatin is completely dissolved.

Stir the seeds, fruit, and nuts into the bowl with the gelatin, then add the peanut-butter mixture. Mix well with a wooden spoon to be sure all the nuts and seeds are well coated and all the liquid is absorbed, then pour into the plastic jar. Put the jar in the refrigerator overnight to set. The next day, invert the jar and gently tap it on the countertop to unmold it, or simply cut the jar away from your seed block with scissors. The treat is ready to serve!

FERMENTING FEED

A good-quality, well-balanced chicken feed should be the foundation of your flock's diet. But did you know that if you add water, you can *increase* the nutritional value of your chickens' feed? By fermenting the grains in the feed, you make it easier for your chickens to digest, and you also increase vitamins B, C, and K as well as digestible proteins *and* increase your chickens' water consumption! The process of fermentation also creates lactic acid, which aids in digestive health and promotes good bacteria in the gut. Studies have shown that fermented feed leads to stronger, thicker eggshells and larger eggs. But the best part is that when you ferment your chickens' feed, you'll notice that they will eat less (up to ⅓ less per day!), thereby decreasing your feed cost. Don't worry, chickens eat less fermented feed because they're getting their required nutrients in a smaller amount of feed, not because they don't love it.

WHAT YOU'LL NEED

Note: You can use starter, grower, or layer feed, or even scratch grains. But be aware that whole-grain or cracked-grain feed ferments far better than crumbles or pellets.

1 day's worth of feed

Gallon (or larger) glass or food-grade plastic container, cleaned with hot, soapy water

Water (filtered and dechlorinated)

Cheesecloth

Rubber band

Pour your chickens' daily feed ration into your container, filling the container only about ⅓ full to leave plenty of room for expansion. You can offer your chickens fermented feed daily or on alternate days, switching back and forth between dry and fermented feed. However, for your first batch, start with just a single day of feed.

Pour water over the feed until the container is about ⅔ full. Stir well, cover with the cheesecloth and rubber band, and set the container out of direct sunlight at room temperature. You can place it indoors or outdoors, as long as the weather is mild and the temperature is 60 degrees Fahrenheit or more. (The warmer the mixture gets, the faster it will ferment.) A few times a day, remove the cheesecloth and stir the mixture to oxygenate the solution. (Bad bacteria can grow in anaerobic pockets if you don't mix it well.) Add water, if needed, as the feed swells and absorbs the liquid, to prevent mold—the feed must always stay submerged and will expand to about double its original volume.

In 3 to 4 days, you should see bubbles forming and the mixture should smell slightly yeasty, like sourdough bread. Strain the liquid out and serve the fermented feed to your chickens. Clean the container thoroughly before starting a new batch. (If you want to start another batch right away, save some of the liquid and add it to the new batch. It will help kick-start the new fermentation.)

If mold does start to grow, or if the mixture starts to smell rancid, toss it and start again with a clean container.

Quick Hack

No time to ferment? Pour cold water over your chickens' feed in summer or use warm water in winter. While there's no probiotic benefit, it does make any powders in the feed stick to the larger pieces, resulting in less waste, and it helps keep your chickens more hydrated in the summer. It's amazing how excited they get about this moistened mash!

SPROUTING GRAINS

If you live in a cold climate where chickens have no access to green grass and weeds for much of the year, your chickens will enjoy eating sprouts to give them some of the nutrients that fresh greens provide in the warmer months. Sprouting grains for your chickens in any climate will offer them an inexpensive, nutritious treat with elevated levels of vitamins, minerals, and protein.

WHAT YOU'LL NEED

Note: You can sprout almost any seed, such as broccoli, radish, mung bean, lentil, peas, and more.

2–3 tablespoons whole grains, such as barley, oats, or wheat

Quart-size Mason jar with lid ring

Water (filtered and dechlorinated)

Window screen or rubber-mesh shelf liner trimmed to fit the top of the jar

Add the whole grains or seeds to the Mason jar. Pour cool water over the grains so they're completely submerged, then screw the lid ring of the jar over the screen or shelf liner. Let the mixture sit overnight.

The next morning, drain the grains through the screened lid, rinse them well in cold water, and then drain them again. Place the jar upside down in a bowl or on a dish to allow it to drain well. Cover with a dishtowel or set the jar inside a cabinet. Since seeds and grains normally grow in dirt, they prefer darkness to begin to sprout. Rinse and drain the grains twice a day, placing them back in the upside-down Mason jar setup afterwards. Make sure to drain them well so they aren't sitting in water.

In a few days, you will start to see the grains split open and begin to sprout. Move the jar to a sunny windowsill, and in less than a week you should have sprouts that have started to grow some green leaves. Feed them to your chickens as a nutritious treat, refrigerating any leftovers.

GROWING FODDER

Fodder is a variation of sprouting that basically is the same concept, except that you let the sprouts grow longer, until they are about 2 inches tall. Just like sprouted grains, fodder is nutritionally dense. Offering your chickens fodder in addition to their layer feed will result in reduced feed cost, eggs with more vibrant orange yolks, and improved health.

You'll need the same materials at left, except you'll need more grains or seeds, and you'll replace the Mason jar with a casserole dish.

First, pour a thin layer of grains or seeds into the casserole dish. Pour water over the grains to completely cover them. Let them sit overnight. The next morning, drain the grains using a colander to prevent the grains from washing down the drain, rinse them, and pour them back into

the dish. Set them on the counter out of the sun at room temperature under a dishtowel (or on a pantry shelf or in a cabinet if you have the room). Rinse and drain the grains twice a day, making sure to drain them well so they aren't sitting in water. After the first few days, try not to disturb the grains too much while draining, holding them in the dish with your hand while you carefully drain the water instead of dumping them into a colander. They'll start growing roots, which will intertwine and form a sodlike mat.

Keep rinsing and draining until you have sprouts that are about 2 inches tall (7 to 10 days). Then lift the whole mat out of the dish and feed it to your chickens, or let them eat the fodder right out of the dish.

HOMEMADE APPLE CIDER VINEGAR

Adding apple cider vinegar to your chickens' water a few times a week not only keeps the waterer cleaner, it controls the bacteria both in the water and in their digestive system. (I add 1 tablespoon of apple cider vinegar per gallon to my chickens' water a few times a week for optimal digestive and respiratory health.) The vinegar boosts good bacteria and is thought to combat the coccidia pathogen, which is present in most chicken runs no matter how fastidiously they are cleaned. Apple cider vinegar also aids in respiratory health and makes drinking water more appealing to chickens.

Raw, unpasteurized apple cider vinegar with the live "mother" in it has the most benefits. It's easy to make at home, and since it can be made from apple cores and peels, it's practically free. You'll want to use organic apples if possible for the best results. I like to save the apple scraps when I bake an apple pie. Or you can freeze the scraps as you eat individual apples until you have about a dozen apples' worth saved up.

WHAT YOU'LL NEED

Apple cores and peels from about a dozen apples

Water (filtered and dechlorinated)

Sugar (optional)

Mason jars with lid rings and lids

Cheesecloth

Large bowl

Heavy plate

Dishtowel

Place the peels and cores in a large glass or stoneware bowl and cover them with at least an inch of water. You can add up to ¼ cup of white sugar per quart of water to help the fermentation process along.

Cover the bowl with a heavy plate and then drape a clean dishtowel over it. (The apple scraps need to be completely submerged in the water so they don't mold.) Place the bowl in a cool, dark location—65 degrees Fahrenheit or so is a good temperature—and let it sit for a week. The mixture should start to bubble and foam a bit as the yeast starts to grow.

At the end of the week, strain out the apple solids and pour the liquid into sterilized Mason jars, leaving about an inch of headspace. Cover each canning jar with a square of doubled cheesecloth and screw the ring part of the top on over it. (Hang onto the flat parts of the lids; you'll need them later.) For now, the cheesecloth allows the yeast to breathe and the vapors to escape.

Note: If you see spots of black mold on the surface, that's okay. It will occur if the mixture isn't kept cool enough or the solids weren't kept completely submerged. You can just scrape it off, discard it, and continue with the remaining liquid.

Leave the jars for about 6 weeks. A film should start forming on the top. This is the "mother"—which is the culture of beneficial bacteria and acids. At this point, you can open the jars and stir or swirl them so the mother settles on the bottom. More of the culture will grow on top. The liquid should start to get cloudy, and you should detect a slight vinegar smell. After 6 weeks, remove the cheesecloth and screw on the rings and lids of the Mason jars. Your vinegar should keep indefinitely, and the flavor will continue to develop over time. Spoon some of the mother into a bowl if you want to start a new batch of apple cider vinegar.

ICE TREATS

During the summer, cooling treats are important to help your chickens beat the heat. If you're at work all day, freezing a water bottle the night before and putting it in a tub of cold water in the shade will help to keep the water cool for them. However, it's also fun to give your chickens these ice treats! They'll love pecking at the ice to reach the frozen nuggets in the middle.

You don't need a recipe for this one. You can use muffin tins, ice-cube trays, bowls, or loaf pans. Really, anything that you can freeze water in will work. Just add some or all of the following:

» Small vegetables, such as peas and corn, or chopped carrots, cucumbers, or squash all work well.

» Blueberries, cut-up strawberries (or strawberry tops), and other types of fruit are favorites of my flock.

» Rose petals, echinacea, bee balm, nasturtium, and violets are all safe and make for pretty treats.

» Cooling herbs such as mint make the treat even better.

Quick Hack

Don't have the time or inclination to make these ice treats? Just slice some watermelon into cubes and freeze them. Your chickens will appreciate the water-laden, cooling treat on a hot day.

HERBAL TEA

Many of us enjoy a cup of herbal tea, but did you know most chickens do as well? Whether you offer your chickens their tea warm in the winter or chilled in the summer months, not only will they appreciate a change of pace from plain water, but their bodies will appreciate the far-reaching benefits of herbs. That said, always offer plain water alongside the brewed tea until you are sure your chickens are drinking the tea. Water is too important for them to go without for even a day . . . and we all know how suspicious chickens are of anything new! It can also help to keep an eye on which herbs your chickens nibble while they are out free-ranging in your garden and start with those for the tea.

To make a tea, pour boiling water over a generous handful of fresh herbs (or a tablespoon of dried herbs) in a Mason jar or other heat-safe container. Let sit until completely cooled. Strain and serve the liquid at room temperature, slightly warmed, or chilled, depending on the season. Try adding a smashed garlic clove to each jar of tea for some extra immune-system boosting.

TEAS TO TRY

Calcium tea: Chickens need a fair amount of calcium in order to make strong eggshells. If they don't get enough calcium in their diet, they will start to leach it from their bones, which can result in weak, brittle bones and can potentially lead to breaks. A good layer feed offers the majority of calcium that your chickens require, and offering crushed eggshells is another way to supplement calcium, but making sure your chickens get enough is always a good idea. This tea might help if you are dealing with an egg-bound hen that needs more calcium to produce the contractions necessary to lay her egg.

Use one or more of these high-calcium herbs: basil, borage, chervil, dandelion, marjoram, red clover, red raspberry

Protein tea: During the molting season, when your chickens are working hard to grow in new feathers before winter, a bit of additional protein is a good idea. Protein is also needed to by young, growing bodies, so an herbal tea such as this is a good one to offer to your baby chicks.

Use one or more of these herbs that are high in protein: basil, chervil, dill, marjoram, oregano, parsley, spearmint, tarragon

Immunity tea: Often a sick chicken will stop eating, but she will usually continue to drink. A tea that is high in natural bacteria-fighting properties can help to strengthen the immune system and give your ailing hen a hand while she recovers.

Use one or more of these herbs to boost the immune system: chamomile, echinacea, oregano, rosemary, thyme, yarrow

Overall healthy tea: Sometimes you just want to give your chickens a vitamin boost. Or maybe a flock member just isn't herself. Combining a few of the most nutrient-dense herbs into one tea is just the thing. This one is also good for growing chicks, to provide them a well-balanced diet.

Use these herbs that are high in vitamins and nutrients: dandelion, parsley, sage

SHARE YOUR KITCHEN SCRAPS

An easy (and free!) way to give your chickens a wide variety of nutritious treats is to keep a dedicated "for the chickens" bowl on the kitchen counter next to your cutting board when you cook. I remember both my mom and my grandmother doing this, and I've continued the tradition. Our chickens get the tips and ends, peels and leaves, scraps and trimmings from most fruits and vegetables, as well as bread products and some meat.

That's right, chickens are omnivores: they can (and will!) eat almost everything. Of course, the healthier you eat, the healthier they'll be eating when they devour your scraps. Sticking with lean meats, veggies, fruit, and whole grains is best when it comes to sharing with your chickens, but the occasional pie crust, last bite of cheeseburger, leftover Chinese take-out, or just-past-the-date yogurt isn't going to kill them. Don't feed them anything moldy or spoiled though! Also skip over anything that's deep-fried or terribly sugary or salty.

As far as foods to completely steer clear of, those would include alcohol, apple cores, avocados, chocolate, coffee grounds, green tomatoes, cherry, peach, or plum pits, and tea bags. Try to limit asparagus, citrus, dairy, onions, white bread, and white potatoes, since these foods can pose health issues in large amounts.

SHARE THAT COOKING WATER TOO

Don't toss out cooking water from vegetables or hard-boiled eggs. If it's unsalted, it makes a hearty drink for your chickens. Let it cool and serve it to your flock. In the summer, you can also pour it into a loaf pan or similar container and freeze it. Then let your chickens peck at it to cool down. Cooking water is also good for using to water your garden. Veggie water will give your plants a nice drink of phosphorus, nitrogen, and potassium. Water you used to boil eggs is packed with calcium, which is beneficial for both your chickens and your calcium-loving garden crops.

AND EVEN SHARE SOME FAT

Many chicken keepers buy commercial suet cakes and a holder and hang the suet in their coops for their chickens to enjoy on cold winter days. Chickens can handle a bit more fat in the winter, since they burn a considerable number of calories trying to stay warm. Homemade suet is a great way to not only recycle your cooking grease, but also offer your chickens a fun treat. So don't toss out that old saucepan! It can make a wonderful vehicle for homemade suet.

I like to use a saucepan with a hole in the handle for my suet. This makes it easy to hang the pot in the run once it's ready. I simply keep the saucepan in the freezer, and any time I cook meat, I pour the drippings into the saucepan. I also layer the pan drippings with assorted unsalted nuts, seeds, and dried fruit as I add the layers of grease. (I try to avoid using bacon fat due to the high nitrate and salt content, but all other types of cooking grease are fine; bacon drippings are even fine in moderation.)

I continue to fill the pot all through the year and then hang it in the run on cold winter days for my chickens to enjoy. It acts as a bit of a boredom buster as well as a warming winter treat.

KEEP WATER FROM FREEZING IN THE WINTER

Chickens do a decent job of keeping themselves warm in the winter, but they do need access to fresh water, which can be a bit of a challenge in the cold months. Even without electricity near the coop, keeping water from freezing isn't difficult if you use this simple trick.

Grab one of those black rubber tubs that you can find at any feed store. The larger your tub is, the better, because the more surface area it has, the longer it will take to freeze. There's advantage 1. Advantage 2 is that the black rubber absorbs the heat from the sun. For advantage 3, drop a

few table tennis balls into the tub with the water. The slightest breeze will get the balls bobbing and create movement on the surface, which prevents the water from freezing completely.

If you do have electricity in your coop or can run an outdoor extension cord, a heated dog water bowl is the easiest way I've found to provide liquid water all winter long. It's inexpensive and easy to refill and clean. A second bowl is perfect for your chickens' oatmeal! Scoring a slow cooker at a secondhand store or repurposing an old one you have on hand is another great way to provide your chickens warm water through the winter.

KEEP WATER COOL IN THE SUMMER

Chickens need water the most in the hot summer months, but of course they don't like to drink warm water. Most chickens will actually stop drinking if their water gets too warm, which can lead to dehydration and a drop in egg production. Keeping your chickens' water cool is easy—just switch out the warm water for cool water on hot days.

Place the tub in the shadiest spot in your yard—ideally you want it shady all day. Next, freeze water in mini loaf pans or muffin tins overnight, placing the blocks in the water before you leave home in the morning. You can even freeze water in gallon jugs if you'll be gone for a very long time.

If you have an extra-hot climate, consider freezing a whole tub of water (or as large a tub as will fit in your freezer) and setting it out at the beginning of the day.

KEEP FEED FRESH

It's one of the worst feelings to buy a bag of feed and store it until you're ready to open it, only to spill it out through a hole in the bottom as you move it. It's equally bad to open a bag of feed and find clumps in it, indicating that the feed has gotten wet. Mold is a danger to chickens, and mice can transmit disease, so you want to keep your chicken feed free of moisture and rodents. An easy way to avoid both of these problems is to transfer the bags of feed into large galvanized pails with lids as soon as you get them home. This will keep the bugs, mice, and moisture out of your chickens' feed.

SAVE LABEL SLEEVES

When you're transferring food to a dedicated container, make sure you save the label sleeve. Chicken-feed recalls aren't all that common—and they're often for something minor. However, every so often there's a serious feed recall that could affect the health of your flock.

No matter the situation, it's important to know exactly where and when the bag of feed you're feeding your flock was manufactured. The hang tag or label on each bag of feed will list not only the manufacturer of the feed, but also the type of feed, lot, batch, and production date. (Sometimes the date will be on the thin white paper strip stitched into the bottom of the bag. In that case, hang on to that too.)

Try attaching a clear plastic pouch to the side of your feed can to hold the labels. Photo sleeves work well, as do shipping sleeves that have adhesive on the back for sticking to packages. Even a clear baggie taped to the side of your feed can will work. Then you can simply slide the feed label into the pouch. Remove and replace it each time you open a new bag of feed. Now if there's a recall or any issue with the feed, you'll have the label at your fingertips.

ADD A FEW HERB SACHETS

If you want to deter insects and rodents from taking up residence in your chicken feed, buy a few simple sachets or stitch up a few sachets as described on page 102. Fill them with a combination of dried herbs that bugs don't like and toss them into your feed pails or wherever you store your feed. Some good pest-repelling choices include bay leaves, catnip, chamomile, lavender, lemon balm, marigold, peppermint, rosemary, spearmint, and thyme. Use only completely dried herbs in sachets that you place in feed, so you don't run the risk of the herbs molding and contaminating the feed.

CRUSHED EGGSHELLS FOR CALCIUM

Even when chickens are eating a good-quality layer feed, they benefit from a supplemental calcium source. You can either purchase commercial crushed oyster shell or save and crush your chickens' eggshells to feed back to them. My chickens actually prefer the crushed eggshells to oyster shell. (Yes, I've done side-by-side taste tests.) Since it's free and easy to feed them the eggshells, that's what I do.

I don't believe that feeding chickens eggshells will lead to unauthorized egg eating. In all the years I've been feeding the eggshells, I've not ever had any issue with that. I personally believe that by providing your chickens plenty of calcium, they won't go looking elsewhere for it—and start eating their own eggs. I do crush up the shells fairly small. You don't want them ground into powder, but maybe ¼ inch in size or so is a good rule of thumb.

Want to give it a try? Here's what you do: Once you've cracked the eggshells and made breakfast, rinse the shell halves under running water and remove the membranes. You can also leave them, but I find the shells dry faster and crush better with the membranes removed. Then set the shells onto a paper towel on the counter to dry. Once they're dry, collect them in a bowl on the counter, crushing them with your fingers every few days to reduce the mass and make more room in the bowl. Then fill your dispenser or bowl for the chickens and let them eat the crushed shells as they choose. Each hen's calcium requirements may be a bit different, so this way each can eat as much or as little as she needs. Young, not-yet-laying pullets won't eat any shells, nor will chicks or roosters. Oh, and see page 41 for some ideas for making your own calcium dispensers!

GRIT FOR DIGESTION

Since chickens don't have teeth, they need help grinding their food. They collect what they eat throughout the day in their crop, where it's stored until they are ready to digest it. Then the food moves to their gizzard, where it's ground up with the help of small stones, pebbles, and coarse dirt that they pick up as they roam their yard. If your chickens can't free-range, it's a good idea to provide them grit, which is available commercially, or you can collect dirt and small stones for them from your property. (Generally, pea-size or smaller grit is what they're looking for.)

One thing I learned the hard way is that if you live in a northern climate, you'll want to collect a few bucketfuls of dirt to use through the winter. The ground will be frozen and possibly covered with snow for part of the year.

CHARCOAL SUPPLEMENT

Grit and calcium are two extremely common supplements for chickens, but charcoal is another that offers amazing benefits for a flock. Animals in the wild have been observed gnawing on charred wood after a forest fire—and with good reason. Charred wood is a good source of potassium, phosphorus, and magnesium. Charcoal also contains calcium, which will result in stronger eggshells when added to your flock's diet. Chickens that nibble on charred

wood have been found to be healthier overall, with better-quality feathers. Charcoal ingested by a chicken also binds with ammonia fumes in its manure and helps to neutralize the fumes, making for a far healthier coop environment.

Providing a piece of charred wood in the coop for your chickens to peck at will allow them to reap the benefits, which include some protection against coccidiosis and other internal parasites. The wood you burn in your fireplace, wood stove, or fire pit (or even in a large metal pail specifically for the chickens) can't have been painted or treated with any toxic chemicals, and clearly you don't want to give them charcoal briquettes. Once you have a few charred pieces of wood, you can leave them in your coop for your chickens to nibble on at their leisure—or you can even nail a piece of wood to your coop wall. Smaller pieces could be offered in a dispenser, as I've shown for the other supplements on the next page.

VINTAGE MATCH-BOX DISPENSER

WHAT YOU'LL NEED

Metal match box

Chalk spray paint

Chalkboard paint

Chalk

Clear polyurethane

Small grit and calcium dispensers are available commercially, or you can just use a small bowl or hanging cage cup, but if you prefer a more rustic or vintage look, scour eBay, antique stores, or yard sales for vintage metal hanging match boxes. You should be able to pick up several for just a few dollars. Don't write off a dented, rusty old box either! A few coats of spray paint (I like to use chalk spray paint for a soft, matte finish) will be all it needs. If you're feeling ambitious, paint the front with chalkboard paint and then, once dry, write "Grit" or "Shells" on it with regular white chalk. Then spray it with clear polyurethane to set it. Let dry completely and hang in your coop. The dispenser is the perfect size to keep a ready supply of supplement available for your flock.

Even easier, hang a soup ladle or 2 on your run fencing if you have a covered area, or in your coop from a nail, then fill the bowl of the ladle with eggshell or grit.

IN THE COOP

The chicken coop provides a safe, comfortable spot in which your chickens can lay their eggs and sleep at night. Whether you build your own coop, buy one fully assembled or as a kit, or convert a garden shed or playhouse into a coop, there are lots of opportunities for some handy upgrades that will make your life and your chickens' lives easier and more comfy. Of course functionality and security from predators is the number-one concern when it comes to coops, but once you've accomplished that, it's time to add a little fun and a bit of frill to the coop in the form of curtains, supplement dispensers, natural cleaning sprays, and even aromatic herbs!

Although there really is no one "perfect" coop size, style, or design, there are some parameters or rules of thumb that are pretty standard things to consider when you're choosing or building your coop or trying to figure out how many chickens will fit. You should allow for the following in your coop:

Floor space: 3–5 square feet/hen

Nesting boxes: 1 nesting box for every 3–4 hens

Roosting bars: 8 inches of roosting bar per hen (a 2×4 board with the 4 inches side facing up is perfect)

Quick Hack

Find an old bookshelf or storage shelf and some plastic dishpans that slide onto the shelves—and you've got makeshift nest boxes! Want some other quick ideas? You can use clean half wine barrels, plastic totes, wooden crates set on one side, plastic kitty-litter pails, or even 5-gallon pails with the lids cut into an arc at the bottom for your chickens to nest in.

CONVERT A CHAIR
into a Nesting Area

Some of my chickens love their privacy, while others don't mind laying in full view of everyone! This portable chair nesting area is great if you have 1 hen separated in a different area who needs a place to lay her eggs. It can also be moved outside under the shade of a tree on those summer days when the coop gets too hot during the day for your hens to sit in for long periods of time.

WHAT YOU'LL NEED

Enamelware, rubber, or metal pan

Old chair or bench

Note: If you have an old bench, even better . . . make 2 nesting bins!

Permanent marker

Cordless drill with ½-inch drill bit

Jigsaw

Paint, stain, or polyurethane (optional)

Using the pan as a template, set it in the center of the seat of the chair and trace around the base with a permanent marker. Make a pilot hole with the drill bit to get started. Then, using the jigsaw, cut the seat, following the circle you traced. Enlarge the circle if needed so the pan will sit in the hole, held in place by the lip of the pan. Paint, stain, or polyurethane the chair if desired.

Set it in the corner of your coop and fill the pan with chopped straw or pine shavings.

HERBS IN NESTING BOXES

Anyone who knows me knows that I love talking about herbs! Studies have shown that wild birds sometimes line their nests with wildflowers, herbs, and weeds. The thought is that the birds have some sense that these plants are beneficial to their babies once they hatch and also help protect the nest from bugs. I love the idea of taking a cue from these birds and adding fresh or dried herbs to my chickens' nesting areas. It looks pretty, smells nice, and just might calm my sitting and laying hens

while keeping bugs out. Choosing aromatic herbs with insect-repelling qualities is the goal here, so some of my favorites to add to the nest boxes are bee balm, catmint, catnip, chamomile, lavender, lemon balm, marigold, mint, pineapple sage, rose petals, rosemary, thyme, and yarrow. They're all perfectly safe and edible for the chickens in case they decide to nibble on some. I add them fresh to the boxes during the growing season and dried during the off-season.

STONE "EGGS"

Egg eating is a bad habit that chickens sometimes get into—usually by accident, after an egg breaks in a nesting box. The chickens get curious and discover how delicious eggs taste! To try and break the habit, some people put golf balls or plastic Easter eggs filled with sand in the nest boxes.

Along the same line of thinking, why not pull some egg-shaped stones from your yard to put in the boxes? The stones have other uses as well. The fake eggs can also be put in a nest to try to encourage one of your chickens to go broody and sit on eggs to hatch them. Also, if you're new to chicken keeping and have young layers, putting the stones in the nesting boxes can show them where they are supposed to lay *their* eggs.

Quick Hack

Make the stones an attractive coop feature by painting them in soft pastel colors! Chickens see in color and even more colors than humans can. However, I've found they don't care what color the fake eggs are, so feel free to paint away. Use nontoxic paint.

Lining your nesting boxes with a piece of rubber shelf liner or piece of old yoga mat can help prevent breakage also. Just cut the piece to fit, then add your favorite box nesting material, i.e., pine shavings, straw, or even dried leaves.

NESTING-BOX BLOCKER

The nest boxes in your coop should be reserved for egg laying. Period. If your chickens get in the habit of sleeping in the boxes, you'll be cleaning them and refreshing the bedding constantly, and you'll likely end up with dirty eggs. The easiest way to break the bad habit of sleeping in the nest boxes is to block them in late afternoon, after everyone is done laying. Hammer a roofing nail or similar nail with a nice large, flat head into each of the 4 corners of your row of boxes. Then cut a piece of chicken wire to fit across the front of the boxes. Attach the wire to the nails, and voilà! The boxes are inaccessible until you remove the wire the following morning to allow egg laying.

NESTING-BOX CURTAINS

Nesting-box curtains have all sorts of uses. Old-time farmers would simply hang old burlap feed bags over their nesting boxes. If you have a nice-looking coop, however, why not make some nicer-looking curtains? You can sew curtains and hang them from a curtain rod, attach them with Velcro, or, even easier, just nail or staple them to the walls of the coop. When they get dusty and are in need of cleaning, just take them down and launder them—or make more. You can make tiebacks so the boxes are only partially blocked, or make basic drape-style curtains. Or you can stick with a simple valance if you're only interested in the cute factor.

You can use fabric you have on hand or buy inexpensive bedsheets or pillowcases at a thrift shop to make into curtains. Old tea towels also make pretty curtains. I use tea towels in my coop and just nail them up, so they're easy to remove when they get soiled.

NESTING-BOX CURTAINS GENERALLY HELP:

» Encourage reluctant layers to use the boxes, since chickens prefer dark, secluded spots

» Encourage a broody hen to sit on eggs by giving her more privacy

» Help a broody hen keep her chicks warmer after they have hatched

» Discourage broodiness from spreading to the rest of the flock since a broody hen stays out of sight

» Prevent egg eating, since eggs are out of view once laid

» Prevent pecking of a laying hen's vent, which can happen if others spy the red, swollen vent as the egg pops out

» Delay eggs from freezing and cracking in the winter by retaining the laying hens' body heat in the box for longer

NO-SEW WINTER
Coop Window Curtains

Just as curtains or drapes help to keep your own home cool in the summer by blocking the sun's rays, and warmer in the winter by reducing the amount of heat that escapes, hanging curtains in your chicken coop can do the same. Drawing the drapes on a hot summer day can shield the inside of your coop from the sun's rays and, even more importantly, will help keep your coop warmer in the winter by preventing your chickens' body heat from escaping through the windows.

And the best part? Even if you don't sew, you can quickly hang pretty curtains that are both functional and attractive. Using a pretty wool plaid shirt will add some seasonal decor to your coop while creating an insulating block over your coop windows. Tie the drapes back during the day, then let them hang down, covering the windows, at night. By using an old shirt (either raid your closet or pick one up cheap at a secondhand store), you can create panels that don't even require any sewing! At the end of the season, take the drapes down and launder them if you wish—and you can replace them with a summer version made from a cotton shirt!

WHAT YOU NEED:

Large or extra-large flannel wool shirt for winter or cotton shirt for summer

Scissors

Hammer

Small nails

Scrap of ribbon

With scissors, cut out the 2 front panels of the shirt into rectangles by cutting down the side seams and across the top under the arms. With the shirt front buttoned, nail the panels across your coop window so the fabric covers the window opening. Unbutton the buttons and tie them back with the ribbon (or you can remove the sleeve cuffs and use them as tiebacks instead of the ribbon), nailing the tiebacks to the wall to secure each panel. Leave the panels hanging down at night as a barrier against the winter cold and pull them back each morning to allow in some sun to warm the coop through the day.

POP-DOOR DRAFT DRAPES

During the winter, partially blocking the little pop door that the chickens use during the day to come in and out of the coop can help keep drafts out, keeping the coop a bit warmer. You can use tea towels, scraps of fabric, or even an old shower curtain cut into strips and hung over the pop-door opening (picture chickens going through a car wash). It might take some time for your chickens to get used to walking through the drapes, but if you get the curtains hung early enough in the fall (and tie them back at first) to let your chickens get accustomed to them, by winter you should be able to let the curtains hang down. They'll keep the door blocked from gusts of wind while still allowing your flock access in and out of the coop.

I used 2 curtain brackets, a short piece of dowel, and a tea towel to make the pop-door drapes shown here. I turned over an inch or so of the top of the towel, sewed a pocket for the dowel to slide through, and then made a cut up the center of the towel. That way I could tie back each panel with a short piece of ribbon I threaded through a U-shaped nail hammered into the coop wall.

NATURAL ANTIBACTERIAL
Hand Gel for the Coop

Biosecurity is important when you raise chickens, for you and for your chickens. Most chicken runs contain traces of salmonella, *E. coli*, and even staphylococcus germs, after all. While this antibacterial hand gel isn't a substitute for a good washing, it is a good practice to squirt some on your hands each time you or your visitors leave the coop. Mount it right by the coop door so you remember.

The aloe vera gel is good for cuts and scrapes, as well as for soothing burns, and the witch hazel is an antimicrobial that will ensure your hands dry quickly. A bit of vitamin E works as a preservative for your gel and also adds a moisturizing property to prevent your hands from drying out. I have chosen some essential oils to use in the gel that offer protection from the specific pathogens that tend to be found in and around chicken areas, as well as general antibacterial protection.

WHAT YOU'LL NEED FOR THE GEL

6 ounces aloe vera gel

6 ounces witch hazel

1 tablespoon liquid vitamin E

30–40 drops of essential oils of your choice (choose 1 or more from the list at right)

» Basil essential oil—protects against *E. coli*

» Cinnamon essential oil—protects against *E. coli* and staph

» Lavender essential oil—protects against staph

» Lemongrass essential oil—protects against staph

» Peppermint essential oil—protects against salmonella and *E. coli*

» Sage essential oil—protects against salmonella

» Thyme essential oil—protects against staph

Note: Unfortunately, these aren't the best-smelling essential oils, but I have been using a blend of lavender, cinnamon, peppermint, thyme, and sage—a little heavier on the cinnamon than the other oils—and it's not terrible.

WHAT YOU'LL NEED FOR THE PUMP

Mason jar with a pump-top attachment

Cordless drill with small drill bit and screw bit

Hose clamp

12-inch piece of 1×6 lumber, stained or painted (optional)

2 short screws

Flathead screwdriver

To make the gel, it's as simple as whisking the ingredients together and then pouring them into the Mason jar. Screw on the pump top and you're ready to go! Squirt your hands liberally with the gel each time you leave your chicken coop. Rub into your skin and let dry. The gel should last for several months. You might want to shake it up every once in a while to redistribute the essential oils.

If you'd like to make the hanging dispenser shown here, use a drill to make 2 pilot holes in the back of the hose clamp, about an inch apart. Then position the jar in the center of the board and screw the clamp into position (or right to your coop wall, if you choose).

Slide the jar into the clamp. Using the screwdriver, tighten the clamp to secure the Mason jar in place.

COOP CLEANER, DEODORIZER, and Pest-Repellent Sprays

Cleaners such as bleach and ammonia are not safe to use around your chickens, but fortunately, there are nontoxic cleaners you can use to clean your coop and to disinfect your feeders and waterers. At the first scent of ammonia in your coop, the bedding should all be raked out and replaced or an ammonia-eliminating product sprinkled on the floor. Between larger cleanings, I like to spray a natural herbal spray in my coop to refresh the air and also to act as a bit of an insect and mouse repellent. Most bugs and rodents don't enjoy the scent of aromatic herbs. So while we enjoy them (and the chickens, who don't have such a great sense of smell anyway, don't seem to mind them), spraying an herbal solution such as this can help keep pests at bay while adding a bit of aromatherapy to the coop. I like to use white vinegar as the base for whichever scent of spray I'm making. It's inexpensive and works well when it comes to knocking down bacteria and mold.

LEMON THYME SPRAY

This bright-yellow spray will add a cheerful citrus scent to your coop. Lemon oil is a natural insect repellent, so it's a nice addition to the herbs in this spray. Thyme is an antiparasitic that helps repel flies.

Here's how you make it: To a quart-size Mason jar, add the peels of 1 lemon, a few sprigs of fresh thyme, and white vinegar to cover the solids. Screw on the jar cover and ring. Let the jar sit for a few weeks, shaking the contents every few days, until the mixture is fragrant and the vinegar scent is gone. Strain into a squirt bottle and squirt around your coop and on the roosting bars as needed.

Other Sprays I Like

Orange peel, cinnamon, and vanilla: This is an original scent that I developed and have been using in my coop for a few years now. I love how the ingredients blend together and smell so good. In fact, I like it so much, I also use the spray to clean my kitchen! With the same process as above, use the peels of 1 orange, a couple of cinnamon sticks, and a couple of fresh vanilla beans.

Lime, lavender, and mint: If mice are a problem in your coop, in addition to moving the feed and water outside, you might try this spray inside the coop. Mice supposedly aren't too fond of mint, so add a healthy handful along with the peels of a lime and a few sprigs of lavender. Again, use the same process as the lemon thyme spray.

Cinnamon pine: Both pine scent and cinnamon can help deter insects, so if herbs aren't your thing, try this variation. Make sure you choose an evergreen variety with a nice piney smell for the most aromatic spray. Cut a few branches, strip off the needles, and crush or chop them a bit to release the pine oils, then add a few more small branches into the jar as well.

VINEGAR CLEANER FOR FEEDERS AND WATERERS

White vinegar, baking soda, and sunlight can all do an amazing job of killing bacteria, mold, and pathogens. Ultraviolet rays are also a powerful disinfectant, and vinegar kills many types of mold, as does baking soda. Using all in combination will keep your chicken feeders and waterers clean and your chickens healthy.

Mix 1 part white vinegar with 1 part water. Using a stiff brush, scrub the inside of the feeders with the solution, then follow up with a sprinkle of baking soda. Add more water and vinegar solution, and keep scrubbing. Finally, rinse the feeders off with a hose and let them dry in direct sun.

Quick Hack

Substitute vodka for the vinegar to make a spray that is ready in a shorter time. (You don't have to wait for the vinegar smell to dissipate.) A few days, shaking it up a couple of times a day, should do it.

CHEMICAL-FREE GLASS CLEANER

While I don't expect my coop to be as clean as my house, I do like to wash the windows in my coop from time to time so my chickens can see out of them! I also have a mirror in my coop that needs to be cleaned from time to time. What I don't want to use in my coop (or my house for that matter!) are toxic, chemical-laden, ammonia-based cleaners. With this supersimple recipe, it's easy and inexpensive to whip up a batch of homemade glass cleaner. Not only does it work, it smells fresh and is perfectly safe to use around your children and pets (although I wouldn't let them drink it because of the alcohol content). The vodka seems to cancel out the strong vinegar smell, but I still like to add a few drops of lemon or other citrus essential oil—I always associate citrus with cleanliness.

WHAT YOU'LL NEED

1 cup water

½ cup white vinegar

½ cup vodka (no need to use the good stuff)

10 drops lemon essential oil

16-ounce squirt bottle

Add the water, vinegar, vodka, and essential oil to the squirt bottle and shake to mix well. Squirt on windows or other glass surfaces and wipe clean with a piece of crumpled newspaper until the glass is dry. Use a fresh sheet of paper if needed to remove any remaining streaks.

HERBAL FLY SPRAY

Many insect sprays can be harmful to both your family and your chickens. However, bugs can transmit disease and are a general nuisance in and around a chicken coop, so you'll want to keep them at bay. Instead of reaching for the toxic chemical sprays, mix up a batch of all-natural herbal fly spray and spritz it around the coop, run, or patio—wherever flies are a problem.

WHAT YOU'LL NEED

4 ounces water

Small pot

Large handful fresh basil, lavender, mint, rosemary, and/or thyme

Pint-size Mason jar with a spray top (or spray bottle of your choice)

4 ounces vodka

Vanilla bean, split

Bring the 4 ounces of water to a boil in the pot and then remove from the burner. Add the handful of herbs and let steep until the water cools to room temperature, then squeeze the herbs to wring out as much liquid as possible. Discard the solids.

Pour the infused water into the Mason jar, add the vodka and the vanilla bean, shake well, and fasten on the spray top. That's it!

THE DEEP LITTER METHOD

I'm a big fan of old-timers' tips and tricks. The deep litter method is one of the first hacks I adopted from my grandparents when my husband and I first started raising chickens. In short, deep litter lets you compost right inside your coop all winter long.

If you garden, you already know that a good compost pile contains both green and brown components. It needs turning to keep it oxygenated, creates heat and good microbes, shouldn't smell, and results in beautiful compost for your garden in early spring. That concept is replicated by turning your whole coop into a composter! You keep adding bedding (whether it be straw or shavings, dried leaves, or pine needles) and turning it (or letting the chickens turn it by tossing in scratch grains or sunflower seeds for them).

It's best to start in the fall and continue through the winter. Begin with a 6-inch layer of bedding on the coop floor. Each morning, you'll want to turn it over to add oxygen and so the manure drops to the bottom. Add a bit more bedding each morning and turn until you have a layer of bedding about 1 foot deep. As it starts to compress and decompose, keep adding more and turning. The chicken manure is your nitrogen, the bedding is the carbon, and the turning adds oxygen.

If done correctly and if your coop is well ventilated, you shouldn't smell manure or ammonia. The decomposition will create a bit of natural heat for your coop, and you will notice that as the winter progresses, you'll start to see fine dry dirt at the bottom of the pile. In the spring, you can rake out all the litter and use it to fertilize your garden soil.

Quick Hack

Straw can be expensive in some areas of the country. Instead, you can use dried leaves, pine needles, or even dried grass clippings in your coop—or a combination thereof—for great (free!) bedding.

STRAW-BALE INSULATION

Straw on the floor of your coop is probably the best, most insulating bedding you can choose for the cold winter months. To insulate your coop even further, stack bales along the walls inside the coop for natural insulation. Far safer than a heat lamp, straw bales will help retain the warmth from the chickens' body heat through the night. A straw bale has an R-value of 25 to 30, which is comparable to a traditional insulated home. (That's why some people choose to build straw-bale homes!)

Once spring comes, there's no need to toss out the bales. You can use the straw as bedding in the coop through the spring and summer, or as garden mulch. You can even make straw-bale cold frames to harden off your seedlings. (See page 160 for more on this).

CHEERFUL COOP LIGHTING

Chickens naturally slow down or stop laying altogether when the days get shorter, but some people add artificial light—giving them the 16 hours or so of light that they require can force them to keep laying.

I prefer to give my chickens the natural break that their bodies must be craving after a rough molting season. That said, sometimes you need a bit of light to see in the coop—to do a headcount, collect eggs, or just say good night! Instead of wiring your coop for electricity or having to remember to bring a flashlight with you to the coop at night, try stringing up some battery-powered twinkle lights. They give off enough light so you can see and also add a bit of cheer year-round! You can usually find a good deal on them after the winter holidays.

Quick Hack

If you live somewhere that gets a lot of snow, don't be so quick to clear the snow off the coop roof or from around the sides. Snow is a really great insulator. (Think igloos!) Leaving snowbanks against your coop and some snow on the roof (assuming the roof can handle the weight) is a neat and easy way to keep your coop insulated through the cold months.

FACE TIME

Did you know that adult hens will use a mirror hanging in the coop to admire themselves? Hang the mirror at chicken eye level and watch as your hens preen and gaze at themselves in the reflection.

If you have a rooster, he might feel threatened by the big, handsome fellow in the mirror. To keep the risk of broken mirrors at a minimum, it's best to hang a mirror only if you have an all-female flock.

JUST-FOR-FUN
Coop Sign

Metal tart pans come in all different shapes and sizes and are readily available at your local thrift shop or secondhand store—or even in your kitchen cabinet. It's simple to turn one into a cute sign for your coop. Keep track of the supplies you need to replenish, track your eggs for the week, or even count down the days until your eggs will hatch using this chalkboard coop sign.

WHAT YOU'LL NEED

Chalkboard paint

Bristled or foam paintbrush

Metal tart pan with removable bottom

Short piece of ribbon or twine

Adhesive (superglue or rubber cement)

Piece of chalk

Paint the bottom of the pan with 2 coats of chalkboard paint, letting it dry completely between coats. While the paint is drying, glue 2 lengths of ribbon or twine to the underside of the ring part of the pan.

Once the paint is dry, glue the bottom in place into the ring, tie the ends of the ribbon into a bow, and hang your sign in your coop. Use chalk to write your shopping list or an inspirational quote for your flock!

THE RUN

From swings to outdoor perches to creative ways to offer treats to your chickens, the run offers lots of opportunities to create a veritable playground for your flock. The more you give them to do, the less you'll have to worry about pecking and other potential issues that can occur when chickens get bored. If you can't free-range because you're not home all day or there's a threat of predators, then some run upgrades are a must. But keep in mind that many of these ideas can be used even with free-ranging chickens.

CHICKEN SWING

Chickens love to swing! They instinctively seek high ground and love to perch. A swing is a great way to keep your chickens from getting bored and the perfect way to allow a hen that's getting picked on to get away from the bully. You might need to install the swing fairly close to the ground at first, until your chickens get used to it. You should eventually position it up above the head level of the chickens, however, so no one gets hit in the head upon dismount!

All you need is a tree with a nice sturdy branch, at least 2 inches in diameter and about 2 feet long. Cut the branch with a handsaw (or carefully with a chainsaw!). Knot a piece of clothesline to each end of the branch and then tie the free ends of the clotheslines to large eyehooks screwed into the top support boards of your run.

EVERGREEN SHELTER

If you have some evergreen trees on your property, you can make a cute pine teepee-like shelter for your flock. It will make a nice spot for them to take a nap or seek respite from the wind. Cut a few branches that are about the same length (maybe 3 to 4 feet long). Using sturdy twine, wrap the branch ends together and secure them. Set the shelter in a nice sunny spot and fan out the needle ends of all the branches so your teepee stands up on its own. Add some smaller branches to cover any holes or thin spots by pushing the ends of the branches up underneath the twine. Add bedding—pine needles, dried leaves, and straw all work well—and watch your chickens enjoy their hideout.

Quick Hack

If you celebrate Christmas with a real tree, after the holidays lean it against the run fencing or lay it down in your run. Be sure to remove all ornaments, hangers, tinsel, and so on, and make sure that the tree wasn't sprayed with any chemicals or pesticides. Your chickens will love to perch in the branches or nap underneath it.

DIY WATER HOOKUP

Providing fresh water for your chickens is so important. Since an egg is made up largely of water, chickens that don't have constant access to fresh water will almost immediately stop or slow down laying eggs. This hack is an easy way to ensure that clean water is just a twist away. By running a hose hookup right to your run, you can easily refill your chickens' water without hauling heavy buckets. Adding a shut-off valve makes the whole thing even easier. You'll soon enjoy the ease of rinsing and cleaning the water bowl, which can be done right there as well. (Just add a hanging scrub brush.)

WHAT YOU'LL NEED

Large metal bowl

Spigot with shut-off valve

Hose

Eyehook, nail, or screw

Stiff brush

Inside your run, where you plan on setting up the feed and water, set the bowl on the ground at the base of a fencepost. Screw the spigot to the fencepost a few inches above the top of the bowl and attach your hose. Screw in the eyehook, nail, or screw to use as a hanger for your scrub brush. That's it! If you're ambitious, you can dig a trench and run PVC pipe for water (or to protect a hose you feed through the pipe), but for many a simple hose run aboveground from the house spigot will work just fine.

HANGING TREAT LOG

Unfortunately, bored chickens often start pecking at each other, eating their own eggs, or otherwise behaving badly. Making this simple hanging treat log can go a long way toward keeping your chickens busy.

WHAT YOU'LL NEED

Cordless drill

Small drill bit (the size of your eyehook)

1½-inch round drill bit

Short piece of log

Large eyehook

Peanut butter or coconut oil

Seeds, nuts, or dried fruit

Twine, chain, or clothesline

Predrill a hole with the small drill bit in one end of the log and screw in the eyehook. Then, using the 1½-inch round drill bit, drill holes about 1 inch into the log (they don't need to go all the way through) in various random spots. Fill each hole with peanut butter or coconut oil mixed with seeds and nuts. If you do decide to drill some of the holes all the way through the log, you can put things such as carrots, long cucumber spears, or other veggies in the holder as well. Note that in warm climates or during the summer, coconut oil should not be used because it will melt.

Use the twine, chain, or clothesline to hang the log in your coop or run to give your chickens a fun, moving treat station! It should keep them busy and out of trouble for at least a little while. Refill the holes as desired.

HANGING TREAT BASKETS

One of my best dump finds was a metal shepherd's crook. I brought it home and sank it into the ground in my run to secure it, then hung a metal basket from each arm. The baskets are perfect for serving treats such as leafy greens, cabbage or lettuce heads, dandelion greens, and other yard weeds.

Quick Hack

Thread a piece of clothesline through the center hole of a metal Bundt pan, knot it underneath to secure it, then hang the pan in your run and fill it with your flock's favorite treats!

TIERED TREAT STATION

Be creative with other things you find around your house! I think at some point everyone buys one of those tiered metal baskets. You put it on your kitchen counter, and it holds bananas that end up turning brown or onions that sprout before you use them. Mine went from my kitchen counter to the laundry room, where it was stuffed with boxes of dryer sheets, wool dryer balls, and other sundries. I dusted it off and turned it into a treat station for the chickens! I like to fill it with all my chickens' favorite treats from the garden and let them feast.

DUST-BATH HERBAL SPA

Chickens dust bathe to keep their feathers clean, to remove excess oils, and to help control parasites such as mites, lice, and ticks. Although your chickens will often find their own spot to bathe in, it's always a good idea to set up a dedicated area for them. A mixture of dry dirt and sand is a good base for your dust bath. Wood ash from your wood stove, fireplace, or fire pit makes a nice addition to the dust bath as well, since charcoal works as a detoxifier and laxative while providing vitamin K (a blood-clotting agent) and calcium when your chickens eat it. You'll notice that as your chickens bathe, they peck at the dirt and also eat small stones and pebbles, which work as grit to help them digest their food.

I started to notice early on in my chicken-keeping journey that my chickens seemed to like to take dust baths in my herb garden. Once I started studying herbs, I believed that my chickens were deliberately choosing which herbs to bathe in. The thyme, rosemary, and lavender seemed to be my chickens' favorite beds for their baths. Coincidentally, those herbs in particular all have insect-repelling properties. I think as the chickens bathed and rubbed against the plants, they were also allowing the essential oils to be absorbed into their skin and rubbed on their feathers, providing them enhanced protection from external parasites. Since many of the culinary herbs come from the Mediterranean and prefer growing in well-draining, dry, sandy conditions, I decided to plant a dust-bath garden for my chickens.

WHAT YOU'LL NEED

Stumps or large rocks

Dirt

Sand

Wood ash

Diatomaceous earth (optional)

Lavender plants

Thyme plants

Rosemary plants

Sage plants

Start by making a ring with your stumps or rocks, at least 4 feet in diameter. You want enough room inside the ring so that several chickens can bathe at once. Dust bathing is definitely a social event for chickens! Fill inside the ring with a mixture of dirt, sand, and wood ash. I use mostly dirt since we have plenty of that, then add enough sand to make a nice dry bathing area. I add a little wood ash when we have it. You can also mix in a bit of diatomaceous earth (DE). Since it's a fine powder like the wood ash, you'll want to mix it into the dirt really well to prevent the powder blowing around in the wind or becoming airborne when your chickens wriggle around in it, possibly causing respiratory issues. If you feel uncomfortable using the DE, then just stick with the dirt and sand and reserve the DE for the times you have a confirmed mite issue (see page 117).

Plant your herbs around the inside perimeter, right up against your stumps or rocks. This will prevent your chickens from scratching up the roots for the most part, but you will likely have to replant the herbs from time to time after a vigorous round of bathing! Fortunately all of these herbs are pretty rugged and forgiving. Water the herbs as needed until they get established. After that, they should be fine without regular watering, unless you live in an extremely dry area.

If your chickens completely ignore this beautiful oasis you have created for them and continue to bathe elsewhere, well, you've just planted the beginning of a pretty herb garden for your yard.

REPELLING MICE

Mice purportedly don't like the scent of mint, so planting lots of mint around your coop and run, or planting it in window boxes, can help keep mice out of your coop. Any variety will do, but peppermint is supposed to be the best. Strewing fresh or dried mint leaves in the nesting boxes and corners of the coop where mice might make nests can help as well. Peppermint and clove essential oils can also help keep mice away. Never use mothballs, rat poison, or any other toxins around your chickens, other pets, or kids.

A simple way to keep your feed away from rodents is to either set the whole feed bag inside a large galvanized trashcan with a lid or pour your feed into a metal trashcan. I caution against feeding or watering your chickens inside the coop. Feed will just attract mice and flies, and water spills and makes a wet mess. Instead, year-round, I feed my chickens outdoors. Using an enamelware roasting pan for feed out in the run allows you to simply place the cover on top at dusk when you lock up your coop, thereby keeping any leftover feed safe from mice and other rodents overnight.

REPELLING SNAKES

If you can get rid of any mice hanging around your coop, you will most likely get rid of any snakes as well, since you'll be removing their main food source—though you may still find the occasional snake that likes to eat eggs. So, first things first: be sure to collect the eggs often. Whether the snakes in your area are venomous or not, you don't want to reach in and find one curled up in one of your nesting boxes!

One of the traditional ways to keep snakes away is with mothballs. However, please *do not* use mothballs anywhere outside, near your chickens, or near other pets. They are extremely toxic; even the fumes are toxic to humans and animals. My advice is to throw away your mothballs (wrapped tightly and disposed of)—or better yet, don't even buy them in the first place!

A better way to solve the problem is to figure out how snakes are getting into the coop and prevent their access. Snakes often use mouse tunnels underground, so start by blocking any tunnels you see with some rocks. Trimming the grass and underbrush around your run will deprive snakes of a place to hide. Sprinkling sulfur powder with cayenne powder mixed in around the perimeter of your coop and run can help as well. (Snakes supposedly don't like the smell of either.) Finally, using ½-inch welded wire along the bottom part of your chicken run can deter snakes as well.

Quick Hack

According to folklore, cowboys used to coil their lassos around them when they slept on the ground at night to keep snakes away. Apparently snakes don't like to slither across a rough surface, or one that looks like another snake. With this in mind, it won't hurt to try running some rough rope across your coop threshold. I've also used a rubber garden hose running around the perimeter of the run, hoping snakes will think it's another snake and steer clear.

WIND AND SUN BLOCK

Old shower curtains or painter's tarps can make a wonderful temporary wind and sun block. In the wintertime, wrap them around a sunny corner of your run to block the winter winds and provide a sheltered area outdoors for your chickens to enjoy some sunshine and fresh air. In the summer, drape them across the top and down 1 side of your run fencing to shield a section from the sun and keep the area dry when it rains. During the cold winter months, I like using clear plastic tarps or shower curtains, because they still let the light in and create a sort of greenhouse for the chickens.

Bales of straw also can make nice wind blocks for your chickens in the winter.

FLYTRAP PAIL

When you clean your coop or dropping board, collect the manure in a pail. Add some spilled chicken feed if there's some on the ground around your feeders. Sprinkle some food-grade diatomaceous earth on top of the contents in the pail. Then set the pail away from your coop and run. You want it close enough to lure any flies away, but far enough not to stink up your run! I set my bucket around the back of my coop, on the far side from the run, and that seems to work well. The manure and feed will lure the flies, then the DE will kill them—and it will also kill any fly larvae that hatch from eggs they lay in the manure bucket.

I like to use an old metal paint scraper to scrape my roots and dropping board, so if yours is rusty and paint splattered, don't toss it! Save it for coop cleaning days. You can also use a kitchen spatula—just be sure that you devote 1 for use in the coop and never use it in your kitchen after that!

QUICK OUTDOOR PERCHES

Have a downed limb after a storm? Cut the branches to fit into the corner of your run, angling the ends a bit with a miter saw, then screw them diagonally to the run supports. Your chickens will thank you when it's muddy or snowy. (Most chickens will love hopping up onto an outdoor perch so their feet don't get dirty and wet.)

Quick Hack

A board screwed to 2 logs and set in your run makes a simple bench for your chickens to hop onto. Chickens love being up off the ground, and this bench fits the bill—even if it's not as fun as a swing! Even easier, just set some logs or stumps in your run. Another benefit of setting logs in the run is that they offer your chickens a bug and worm buffet. Periodically turn them over and let your chickens gobble up all the insects underneath! (And yes, that's a chicken in a tutu! See page 168 for more details.)

DIY "SCAREHAWK"

Scarecrows have been used by farmers for generations to keep crows from eating their corn and other crops. Yet as a chicken keeper, crows are your friends. Crows will generally scare hawks and other aerial predators away and will alert your chickens to their presence. They generally aren't a problem for chickens, though occasionally they will steal an egg, and you do need to keep an eye on them if you have young chicks.

All that said, you can still use a scarecrow in your yard near your coop to keep hawks and other raptors away and to trick predators such as foxes into thinking there's someone outside with your chickens. (I think of it as a "scarehawk" instead!) Of course, there's no substitute for a dedicated livestock guard animal, enclosed run, or personal supervision when your chickens are free-ranging, but making a scarehawk for your backyard can't hurt. Be sure to move it around every so often to keep the predators on their toes.

WHAT YOU'LL NEED

Assorted clothing items (shirt, skirt, garden gloves, straw hat)

Note: Using old clothing, you can be as creative as you wish. Keep in mind that lighter, thinner fabrics will move more in the wind, and a bunting of cloth scraps will add even more movement to keep predators away.

4-foot length 2×4 lumber for the arms

6-foot length 2×4 lumber for the body

Note: You can also use a 1×4 or 1×2 for both boards, if that's what you have available.

Nails or long screws

Yarn for the hair

2 short pieces of ribbon

Cloth scraps

Clothesline

5-gallon bucket

Sand or rocks

Run the arms of the shirt or dress onto the shorter board first. Then attach that board to the longer board using nails or screws, about 1 foot from the top of the upright longer board, to form a cross shape. You can attach the arms at a slight angle; they don't need to be set perpendicular to the upright board.

Finish dressing your scarehawk as desired, nailing or stapling the clothes in place. Finish up with the yarn hair tied with ribbon, then add the hat and gloves. Cut or rip some colorful pieces of cloth into long strips and tie them at intervals to a length

of clothesline to make a bunting, and then tie that to your scarehawk's hands.

Decide where you want your scarehawk to be—preferably where your chickens typically like to free-range. Then plant it in a 5-gallon bucket and fill the bucket with sand or rocks, wedging in the rocks to keep the scarehawk upright. Move your scarehawk around frequently so predators don't get used to seeing it in the same place . . . or, better yet, make a few scarehawks!

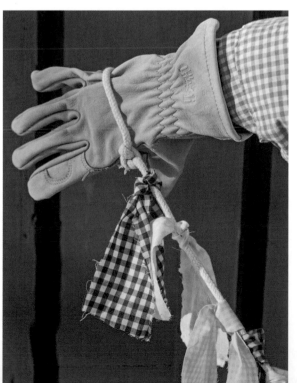

Quick Hack

Another easy and free way to keep predators from lurking around your coop and run is to have the males in your household urinate round the perimeter on a regular basis. Female urine will also deter human-shy predators, but it just seems that guys (and especially little boys) willingly volunteer for the duty—and it's a bit easier for them, when you get right down to it!

CHAPTER

4

RAISING CHICKS ON THE CHEAP

Many of us have heard an urban legend about a kid who hatched chicken eggs in a cardboard box under a desk lamp. However, I have to say that trying to build your own incubator probably isn't the best idea! So much can go wrong if the temperature or humidity isn't spot-on, so it's prudent to spend some money on a good-quality product that ensures the best hatch rate possible.

That said, your own flock can provide the ultimate hack: let your chickens hatch the eggs! It's free and it's easy! It's incredibly reliable: you will experience great hatch rates using chickens to hatch eggs. The best part is since the chicks hatch right in the coop, they are born part of the flock. You avoid all the messy and potentially injury-inducing integration issues that you would face trying to add new flock members. All this saved time should leave you plenty of time to work on fun projects.

This chapter is all about hacks and projects that make your chicks' lives a bit better. From chick-size feeders and waterers crafted from items you have around the house to a simple homemade brooder box and ideas to keep your chicks from getting bored, you'll find plenty to keep you busy!

FEATHER-DUSTER "MOTHER HEN"

If you're feeling bad because your incubator-hatched chicks (brooder babies) don't have a mother, try hanging a feather duster for them in the brooder, suspended just a few inches above the bedding. They will love snuggling underneath it or napping in the darkness it provides. Ask around in your family to see if anyone has a feather duster they're not using or pick one up fairly inexpensively at a home-goods or dollar store.

Quick Hack

Collect feathers from your chickens during the molting season, tie them together with twine at the tips, and make your own feather duster!

MIRROR BOREDOM BUSTER

Hanging a small mirror in the brooder is a great way to keep your chicks busy. They will enjoy standing and preening in the mirror! Don't believe me? Give it a try. A small mirror with a handle is the easiest to simply hang on the side of your brooder.

PAPER-TOWEL-ROLL CANDLER

Let's start at the beginning, before we even hear the first peep. While I don't recommend making your own incubator, making a homemade candler is a simple task. During the incubation period, it's helpful to be able to see inside the eggshell to be sure the embryo is developing correctly. Shining a bright light under the egg in a dark room will allow you to see inside. Before electricity, a candle was used, hence the term *candling*.

Nowadays, commercial gadgets are available, but if you don't already have one, give this a try. Cut a paper-towel-roll to about 4 inches long (or grab a toilet-tissue-roll). Insert a small, bright flashlight through the center and place a fertile egg on top of the roll; you will be able to peer through the shell just fine without juggling the egg.

Note: Be careful not to break the egg, and don't leave it in place too long, as the heat from a flashlight can damage the embryo. A quick look is all it takes.

Early in the incubation period, you will see spider veins. Then, closer to the hatch date, you'll see an enlarged air pocket and a dark blob that is the growing embryo. If you see clear through the shell after about a week, it's likely that egg isn't fertilized. If you see a ring inside the egg, that means that bacteria has gotten inside and the egg is no good. Refer to a good hatching book or educational website for more details on what you will see throughout the incubation period!

THE INCUBATOR SHELF LINER

If you are using an incubator, generally the bottom will be slippery plastic or wire, which is not very comfortable for little feet. Cut a piece of rubber shelf liner to fit the base of your incubator instead. Not only will it give your chicks something to grip onto after they hatch, it will also prevent the eggs from rolling around during the incubation period while you're turning them if your incubator is manual. (If you have an automatic turner tray that you remove for the last 3 days during lockdown, you can put the shelf liner down then.)

THE BROODER SHELF LINER

Don't stop at the incubator! Cut another piece of rubber shelf liner to fit in the bottom of your brooder. Providing a textured surface for your chicks' little feet will help prevent a condition called spraddle leg, in which a chick's legs start to splay or spread—making it difficult or impossible for the chick to walk. Although newspaper is great for absorbing any water that spills, it's too slippery when it gets wet. A piece of shelf liner over a few layers of newspaper makes a safe, inexpensive liner for your brooder. When it needs to be changed out, toss the newspapers, hose off the shelf liner, and reuse it. I like to cut 2 pieces, so when one is drying, I can use the other.

SPRADDLE-LEG TREATMENT

If chicks are brooded on a slippery surface, or if they are subjected to too little moisture or too much heat in the incubator, they can develop a condition called spraddle leg or splayed leg. Their legs start to slide out to the sides, because their young muscles aren't strong enough to keep their legs in place. Caught right away, it's an easy fix. But if left untreated, it can progress to the point where the chick can't stand or walk.

To treat spraddle leg, you can cut a piece of Vetrap long enough (approximately ½×5 inches) to wrap around both legs with space in between the legs. Wrap 1 end around the first leg, then, with the legs in a normal upright position, pull the Vetrap tight and wrap the other end around the second leg, leaving a space between the 2 legs so the chick is standing normally. Use medical tape to secure the Vetrap at both ends and effectively hobble the chick with its legs slightly apart. Tighten the dressing periodically if it becomes loose. The chick will get

Quick Hack

In a pinch, you can try the spraddle-leg treatment with a sturdy Band-Aid, wrapping the sticky ends around each leg with the gauze part in between the legs.

used to walking with the bandage fairly quickly but should be separated from the other chicks if it is repeatedly getting knocked down. Several days with the bandage on should be sufficient for the leg muscles to strengthen. As soon as the chick can walk unaided, you can stop the treatment.

Since spraddle leg can be caused by a vitamin deficiency, supplementing your chicks' diet with some blackstrap molasses or a poultry vitamin supplement can also help.

PASTY BUTT: REMOVE IT AND PREVENT IT

Sometimes a baby chick develops a condition called pasty butt, in which its vent becomes blocked with caked-on poop. It quite literally becomes stopped up. If you don't remove the accumulation, the chick will die. Pasty butt is usually caused by stress or being chilled. (Chicks that are shipped through the mail are more likely to succumb to it than those hatched at home.) It is easily treated.

Here's what to do if you encounter pasty butt: dip a cotton swab in warm water or olive oil, then dab the stuck-on feces gently with the damp swab. Be sure to check at least once or twice a day to be sure the vent is clear. If it's not clear, dab it again with a moistened swab. Try sprinkling some cornmeal or ground raw oats on top of the chick feed. This can help the chick's digestive system clear the stopped-up feces on the inside while you're keeping the feces from continuing to accumulate on the outside.

PLASTIC-TOTE BROODER

A good brooder doesn't need much. It should keep your chicks safe from pets (or curious children), it should be draft free, and you should be able to warm it with a heat lamp or radiant panel heater. It should have enough room for your chicks to move around comfortably and should include space for their feed and water . . . and maybe a few boredom busters too!

I like to use a large plastic tote as a brooder for my chicks. If you're raising just a handful, they can spend at least the first few weeks in the tote. Then, hopefully they can start going outside on nice, sunny days and only sleeping in the brooder.

WHAT YOU'LL NEED

Permanent marker

Ruler

Large clear plastic storage tote with lid

Note: Buy the largest tote you can find at your home-improvement store—the bigger the better (at least 64 quarts for 5–6 chicks).

Box cutter or sharp knife

Wire cutters

Scrap piece of ½-inch hardware cloth, about the same size as the cover of the tote

Cordless drill with drill bit

10 zip ties

Using the permanent marker and ruler, trace around the inner edge of the tote lid, leaving a lip of about 2 inches all the way around. Carefully score the plastic with the box cutter, then cut all the way through to remove the center of the lid.

Using the wire cutters, trim the hardware cloth to a size about 2 inches larger than the window you just cut in the tote lid.

Using the drill, drill 2 holes (about ½ inch apart and 1 inch from the cutout edge) at each corner of the cutout. Then drill 2 pairs of holes along each of the long sides and 1 pair of holes on each of the short sides.

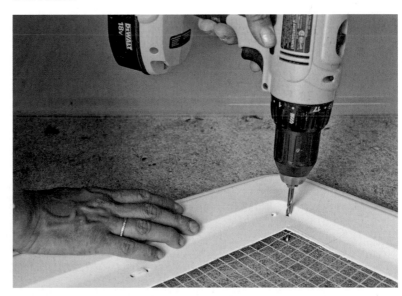

With the tote cover upside down, hold the piece of hardware cloth in place over the cutout and then thread a zip tie through each set of holes and pull to tighten, securing the hardware cloth against the lid.

PLASTIC TOTE FOR OUTDOORS

On nice warm days you can get your chicks outside for a bit, but they need to be kept safe, of course, from dangers and threats such as hawks, dogs, or even cats when the chicks are still small. I like to make a mobile outdoor pen for my chicks. If nothing else, it's a good place to put them when I need to clean their brooder. Setting it in the sun right outside the kitchen door keeps them safe while giving them some fresh air, an opportunity to explore a bit and nibble on grass, plus a bit of new scenery.

Using a box cutter, cut out a window in each of the two long sides of a large plastic storage tote, leaving a 1- or 2-inch edge around the perimeter of the tote. Attach welded wire using zip ties as described for the brooder top on pages 93 and 94. Set the tote upside down in the grass. And voilà! You've got a safe place for your baby chicks to do a bit of exploring out in the yard while you're cleaning.

Quick Hack

You can use a cardboard box, large metal washtub, stock tank, horse trough, puppy playpen, or spare bathtub as a makeshift brooder. Just don't forget to cover the top with a piece of welded wire to prevent your chicks from escaping. (It isn't too long before they can flutter and flap their way out.)

CHICK-SIZE ROOST

Making this chick-size bar for your brooder will help your baby chicks practice. Even chicks just a few days old will have fun hopping up onto the dowel and balancing on it. To make this project as shown, you'll need a dowel and boards cut to specific sizes, but a thin branch and a few pieces of scrap wood can also work for a variation on this idea. Or you can do what I did and repurpose an old clothes-drying rack!

WHAT YOU'LL NEED

Wood glue

Wooden clothes-drying-rack supports or 2 short (5-inch) pieces of wood (for instance, 1×2s with a hole drilled in 1 end for the dowel or branch to slide into)

1×6 or other thin board, 12 inches long

12-inch dowel from the drying rack or branch

Wood screws (optional)

Miter saw or circular saw

Using your saw, cut your board so you have a 12-inch-long board for the base. Cut the ends of your drying-rack supports so you have 2 5-inch-long pieces of wood with a hole each end to use as the supports for the roost. Cut 1 of the drying-rack dowels to 12 inches long.

Glue the short pieces of wood to either end of the longer board, with the ends with the holes in them at the top. Be sure to slide the dowel through the holes before gluing the second piece in place. Let dry completely.

You can secure the wood pieces to the base with wood screws from underneath if desired once the glue is dry.

WARMTH WITHOUT A HEAT LAMP

Quick Hack

Screw a short branch or dowel diagonally into 1 corner of your brooder box, about 4 inches up. You'll need to cut each end at an angle, but that's about all the adjustment it should need. You also might find a branch the perfect size to wedge in the brooder.

If you've ever wondered what people did before the days of heat lamps and ready-made brooders (or what people do in case of a power outage nowadays), listen up. Minnie Rose Lovgreen, who raised chickens for 60 years and was the author of *Recipe for Raising Chickens*, has a suggestion. First, put the chicks in a box with shredded newspaper (straw would work well too). Next, fill a glass Mason jar with hot water, then cap it and cover in aluminum foil before placing it in the box. This will radiate some heat. Then cover the box with an old wool sweater to keep the warmth inside, and place it near the wood stove or radiator. You'll have nice toasty chicks for quite a while and can refresh the Mason jar as needed.

HANGING CHICK FEEDER AND WATERER

Mason-jar feeders and waterers are probably the most common style for baby chicks. I try to avoid using plastics whenever possible, so I prefer the glass jar–metal base variety. Anyone who has raised chicks knows that the feeders and waterers get full of bedding material, dirt, and poop pretty fast if you set them on the ground in the brooder, so here's an easy way to keep them cleaner longer. You just need to hang them!

WHAT YOU'LL NEED

Mason jar with a regular mouth (pint- or quart-size will work)

Epoxy

Small pot cover (just large enough to overhang the bottom of the Mason jar, or about 5 inches in diameter—if you don't have a spare, head to a secondhand shop)

Base from a chick feeder or waterer

Short length of chain

Carabiner

Eyehook

Using epoxy (be generous), glue the pot lid upside down to the bottom of the Mason jar. Set aside to cure.

Once the epoxy is completely dry, flip the jar over and fill it with water or feed and screw on the base. Loop a small chain through the handle of the lid and secure both ends with a carabiner. Hang the jar from a large eyehook screwed into your heat lamp stand crossbar or the top of your brooder. This way you can adjust the chain length to raise the jar as your chicks grow.

CREATIVE FEEDER AND WATERER IDEAS

An empty egg carton makes a simple chick feeder. Simply cut off the top of the carton and fill the bottom with chick feed. Your chicks will enjoy lining up for a quick bite to eat. Have only a few chicks? Cut the bottom in half so you have just 6 feed compartments. Toss the carton when it gets dirty.

A small-terra cotta planter base also makes a nice feeder for just a few chicks. The terra-cotta material will naturally wick away moisture from the feed to keep it dry.

For a waterer, use food-safe, waterproof sealant to glue the ring of a wide-mouth Mason jar to the lid—with the top inverted. Glue another ring to the underside of the first ring for a bit more height.

Let the glue dry completely and then fill it with water. (Don't forget to add a few small stones so your chicks don't accidentally fall into the water and and drown.) Chicks are attracted to shiny things, so the metal and wet stones will encourage them to drink more water. Better yet, make a few for your brooder so if one tips over or gets dirty, your chicks have an alternative source of clean water.

If plastic doesn't bother you, start saving your take-out containers when you're expecting chicks. These containers can be used as feeders, waterers, and even for a small dust bath. If you add apple cider vinegar to your chicks' water a few times a week—I recommend just a few drops to get them used to the taste—you'll want to use plastic water dishes those days. Vinegar will rust metal.

KEY-RACK HERB HACK

Fresh herbs are a wonderful treat for your baby chicks. They're nutritious and packed with vitamins. Nearly all the culinary herbs have great health benefits for chicks (and grown hens too!), but I have found that there are some that chicks prefer. Giving them a selection of popular herbs in the brooder is a great way to keep them busy and also get them used to the taste of the various herbs.

Try using a discarded key rack screwed into the side of your brooder as shown here! It's an easy way to hang bunches of a variety of herbs for the chicks. If you can't screw a rack into your brooder (say, if it's a puppy playpen or metal tub), then you can use string or twine to tie the herbs together into bunches and hang them instead. Just be sure to remove the string when the herbs are gone. As always, whenever you give your chicks anything to eat other than their chick feed, they will need a dish of grit to help them digest the fibrous materials.

BENEFICIAL HERBS FOR CHICKS

» **Herbs for overall health:** parsley and sage

» **Herbs for feather growth:** dill and mint

» **Herbs for respiratory health:** basil and clover

» **Herbs for strong immune systems:** cilantro and oregano

» **Herbs high in protein:** marjoram and tarragon

HERBAL SACHETS

Baby chicks like something soft to sit on or nestle against when they sleep. Wild birds will often line their nests with herbs and wildflowers, presumably for the various benefits of the essential oils as the baby birds rub against them. It's easy to do the same for your baby chicks. Make a simple sachet using a scrap piece of tulle. Fill it with fresh aromatic herbs, then tuck it in the brooder for your baby chicks to snuggle with. Since these are so quick to make, I like to sew a bunch at once, fill them, and then store them in the refrigerator. As one wilts or gets messy, I toss it and replace it with a new one. A sewing machine makes this project a breeze, but you can also sew the sachets by hand if you wish.

WHAT YOU'LL NEED

Tulle, cut to 8 inches square
(1 square for each sachet)

Sewing needle or sewing machine

Matching thread

Assortment of freshly snipped herbs, such as lavender, rosemary, mint, lemon balm, yarrow, thyme, or echinacea

Fold the square of tulle in half. Sew along 1 short end and up the long open side, leaving the other short end open.

Turn the sachet right side out (with the sewn seams on the inside). Fill each bag with herbs, to about ½ inch from the top. Sew along the top edge to close your sachet, then trim the excess thread.

Quick Hack

If you're not a skilled sewer or if you're short on time, you may be able to make an herb sachet with just a couple of hand stitches. Try a stuffed animal, an eye pillow, a small beanbag, or even an old sock. (Remove the stitches of 1 seam, replace the filling with herbs, and stitch the seam back together.)

CLUMPS OF GRASS AND DIRT

We all know the feeling: sometimes your chicks can't be outside, and you want to bring the outdoors inside to them! I love putting clumps of grass and dirt from our yard into the brooder, and I recommend you do the same. (Just be sure it's not treated with any pesticides, herbicides, or chemical fertilizers.) Not only do my baby chicks love to nibble on the ends of the grass and hide behind them, they scratch in the dirt looking for seeds and bugs. They'll also eat some of the dirt, which will double as the grit they need to digest their food. Replace the clumps daily or as they start to dry out.

FEED IN A PINCH

Run out of chick feed? It's best not to feed baby chicks layer feed, because the extra calcium in it can lead to kidney issues later in life. A better idea is to make your chicks some oatmeal—you can even mix in a little bit of cornmeal. Oats added to a chicks' diet have been shown to result in healthier chicks that are more heat tolerant as adult hens. Scrambled eggs are another good stand-in for a day or two, until you can get to the store for more feed. Of course neither oats nor eggs should be fed on a long-term basis, but in a pinch, at least your baby chicks won't go hungry! Both oats and eggs also make nutritious occasional treats for your chicks.

Quick Hack

Fill a small plastic or stoneware container with coarse dirt from your garden or yard. Your baby chicks will love practicing their dirt baths in it. Not only is it a good practice to get into, it helps keep them from getting bored. Bored chicks can start to peck at one another, which is never a good thing.

Disclaimer: Not many studies have been done on using herbs for poultry health care, but I'm convinced that herbal care works. Beyond my own personal experience, many old-timers' methods and anecdotal remedies rely on herbs, and in recent years the studies have started to come out backing up at least some herbal remedies (notably the potential antibiotic properties of oregano in commercial flocks). In the coming years, I'm hopeful that the money will become available so that studies can be done to authenticate many more natural and herbal remedies. Regardless, I stick with culinary herbs, which are safe and edible and won't harm your flock. Thus, at the very least, you'll do no harm.

CHAPTER

5

CHICKEN HEALTH

Keeping your chickens healthy can be challenging, but many natural preventives and remedies have worked for me over the years. Since we eat our chickens' eggs, I don't really feel comfortable treating my chickens with commercial synthetic chemicals and medications (which sometimes dictate a period of time during which you shouldn't eat the eggs that chicken lays). Fortunately, there's rarely any need to if you know a few tricks and believe in herbal medicine.

BASIC FIRST-AID KIT

Any time you raise animals, you should have a first-aid kit close at hand. Many of the common items you likely have in your own first-aid kit are things that will come in handy for treating your chickens. While predator attacks are the most common reason for a chicken to need care, respiratory problems can crop up, as well as digestive problems, impacted crops, or egg binding. Many of these issues are easily treated if caught early and if you have the proper supplies on hand.

It's best to devote a container to your chicken first-aid items, so in the event of an emergency, you're not running around looking for things. A small plastic container with a lid or covered pail makes a perfect kit.

Quick tip: write your vet's phone number and address on the lid of your first-aid kit with a permanent marker, so anyone caring for your flock will have that information at their fingertips in the case of an emergency.

Your chicken first-aid kit should contain at least the following supplies: eyedropper, plastic syringe, gauze pads, Band-Aids, Vetrap, small pair of scissors, scalpel, tweezers, wooden craft sticks, towel, and rubber gloves. It's also a good idea to keep a small dog or cat crate on hand in case you need to immobilize or transport a sick or injured chicken.

Taking a walk down your feed-store aisle will also give you an idea of the many safe and natural products that are a good idea to keep on hand, such as Green Goo (an antiseptic salve), VetRx (for respiratory issues), Nutri-Drench (a nutrient-rich energy supplement), probiotic powder (for digestive issues), saline solution, vitamins and electrolytes, and liquid calcium.

KITCHEN FIRST-AID KIT

While preventing injury and illness whenever possible is always the best course of action, it is important to keep first-aid items on hand. You can and should build a complete first-aid kit with a mix of commercial products and home remedies. However, you can also raid your kitchen in a pinch. At the very least, these items can supplement your normal first-aid kit nicely.

» **Blackstrap molasses**—Packed with nutrients, blackstrap molasses can help an ailing chick or chicken. It can also help flush toxins in the case of accidental poisoning. Offering a small dish of molasses to a hen that is listless, seems to lack energy, or is suffering from stress (for instance, from a predator attack) is recommended. Chickens

generally like the taste of molasses, so free-choice feeding will allow the hen to eat as much or as little as she feels she needs.

» **Blue food coloring**—Apply this to a wounded area or where new feathers are growing to discourage pecking.

» **Cayenne powder**—Sprinkling some cayenne powder into your chickens' feed during the cold months is thought to help prevent frostbite, as it improves circulation.

» **Charcoal**—Activated charcoal has historically been used to flush toxins out of the body. If you think one of your chickens has eaten something toxic, offering a charcoal slushie (activated charcoal powder mixed with water until it's a slushy consistency) can help remove whatever it is she shouldn't have eaten. Charcoal also contains vitamin K, which is a blood-clotting agent, so a bit of charcoal slushie can help a chicken bleeding from internal or external injuries. Studies done in Africa at the Mbarara University of Science and Technology show that wood ash and chili or cayenne pepper are common among local farmers to combat not only salmonella but also Newcastle disease and fowlpox. Like other herbal remedies, however, dosages and length of time to treat are not standardized, so free-choice feeding seems to be the best bet.

» **Cinnamon**—Taken internally, cinnamon can help relieve respiratory distress and is also thought to be an immune-system booster. Sprinkling some cinnamon over your chickens' feed or oatmeal treat is beneficial, especially in the winter, when they may not be out in the fresh air as much. Some cinnamon applied to a wound can help stop bleeding and will also provide antiseptic properties to prevent infection.

» **Coconut oil**—This useful oil helps prevent or treat frostbite when applied to combs or feet, smothers scaly leg mites (see page 120), and can help treat wounds with its antimicrobial properties. Warm it a bit before smearing it onto combs or wattles for an easier application.

» **Cornstarch**—A sprinkling of cornstarch helps stop bleeding fast when applied to a minor wound or cut.

» **Epsom salts**—Epsom salts in a chicken's water can help flush toxins in case of accidental poisoning and can also help resolve impacted crop issues. Use a teaspoon per cup of water, increasing that to a teaspoon per ounce of water for more severe cases.

» **Green tea**—This human favorite is thought to combat coccidiosis when offered to chickens. Brew a cup of strong green tea as you would for yourself, then let it cool to room temperature and offer it free-choice to your flock.

» **Green or black tea bag**—Tea is also known for its ability to stop bleeding. Steep a tea bag in warm water until thoroughly moistened, squeeze it out, and hold it gently against a cut or minor wound to stop blood flow in a pinch.

» **Honey**—When applied to a wound, honey acts as a natural antiseptic to kill bad bacteria and will help heal the injured skin tissue. Substitute honey for an antibacterial ointment in a pinch.

» **Olive oil**—Offer a tablespoon or so of olive oil to a chicken with an impacted crop, or try applying a small amount to the vent of a chicken suffering from egg binding to help things slide easier.

» **Sugar**—Use sugar to help to tighten the vent in the case of a prolapse. Smear the vent with olive oil and then pat some sugar on. It acts almost like a hemorrhoid cream, tightening the skin. Witch hazel will also work in a similar fashion.

» **Vitamin E**—The oil in vitamin E capsules can be applied to a wound to help it heal and prevent infection. Taken internally, vitamin E is also helpful in treating wry neck, a condition that occurs mainly in baby chicks in which they lack the muscles to hold their heads upright.

» **Yogurt**—Offer a small dish of yogurt free-choice to a chicken suffering egg binding, for added calcium. An occasional yogurt treat will also help keep your flock's digestive systems full of good bacteria and help flush out internal parasites.

HOMEMADE HERBAL SALVES

Salves or ointments come in handy for lots of things in chicken keeping. From injuries as the result of a predator attack to flock pecking issues and frostbite, different salves can help with different issues, depending on their ingredients.

WHAT YOU'LL NEED

Handful of fresh herbs

Pint-size Mason jar

1–2 cups olive oil

Cheesecloth

Rubber band

2 tablespoons coconut oil

2 tablespoons beeswax

Measuring cup or other heatproof container

Liquid vitamin E

Add the fresh herbs to the Mason jar. Pour in enough olive oil to cover the herbs completely. Cut a piece of cheesecloth to fit the top of the jar and secure it in place with the rubber band. Infuse the herbs in the oil for 2 to 3 weeks (outside in the sunlight is perfect), shaking the jars every day or so.

After a few weeks of infusion, once your oils have taken on the colors from the herbs, strain the liquids and discard the solids. Melt the coconut oil and beeswax in a glass measuring cup over a double boiler, or in the microwave, until liquid. Stir in ½ cup of the infused oil and a few drops of vitamin E. Let cool to room temperature so the mixture solidifies. Now your salve is ready to use!

Note: to make any of these salves in a hurry, skip the fresh herbs and use several drops of essential oils instead.

Depending on the type of salve you are making, you can choose 1 or several herbs from each of the following categories.

FIRST-AID SALVE

Cuts, scrapes, and wounds from predators (or pecking-order shakeups) are fairly common in chicken keeping. Keeping a general first-aid salve handy to smear on the site of an injury is always a good idea.

Antiseptic herbs: *bee balm, yarrow, calendula, thyme*

SOOTHING ANTI-INFLAMMATORY SALVE

Sprains and strains are common due to rough landings coming down from the roost or the nesting boxes—sometimes even from vigorous mating on the part of a rooster. Treating the injured hen's leg or foot with this salve can help bring the swelling down.

Herbs with anti-inflammatory properties: *plantain, rose petals, chamomile*

BUMBLEFOOT

Bumblefoot is basically a staph infection that begins on the bottom of a chicken's foot due to a small cut that gets infected. You'll need to slice off the black scab that forms at the site of the infection, then remove the hard kernel and white stringy infected material itself from inside. Then you can treat the area with a healing salve before wrapping.

Herbs with the ability to combat a staph infection: *sage, thyme, echinacea, turmeric, yarrow*

FROSTBITE

Exposed combs, wattles, and even feet can succumb to frostbite. Gently smearing a bit of healing salve at the site of the frostbite will not only help heal the affected tissue but also provide antibacterial and anti-inflammatory properties that help reduce pain. This salve can also be used as a preventive and applied to combs prior to a cold snap. The natural oils in the herbs will improve blood circulation and protect the sensitive body parts, preventing further damage.

Herbs to help with frostbite: *chamomile, calendula, lavender, lemon balm, rosemary*

HOMEMADE ELECTROLYTES

When the temperature starts to rise in the summer, your chickens will start to feel it. Considering they're most comfortable in temperatures between 40 and 65 degrees Fahrenheit, chickens can begin to suffer from heat exhaustion when the mercury rises above 82 degrees. With this recipe you can help them beat the summer heat, replacing some of the electrolytes and nutrients they lose when they pant. Electrolytes are also good for new baby chicks (especially if you had them shipped to you) and for chickens recovering from a predator attack, injury, or illness. Electrolytes can be added to the water on hot days using the recipe below.

WHAT YOU'LL NEED

8 teaspoons granulated sugar

½ teaspoon sea salt

½ teaspoon baking soda

½ teaspoon potassium chloride (optional)

For this recipe, I like to mix up a container of the dry ingredients, so that I can just measure out the correct amount as needed and add it to my chickens' water. Just stir or whisk the dry ingredients in a small container with a lid. To use, measure out the dry mixture as follows: 1½ teaspoon to 1 quart of water or 6 teaspoons per 1 gallon of water.

Offer in addition to plain water on hot days. For added relief from the heat, try mixing up the electrolytes in water, then pouring the liquid into ice-cube trays and freezing them. Your chickens will appreciate the frozen treat.

Signs of heat exhaustion include heavy panting, pale combs and wattles, unsteadiness, holding wings out from the body, standing still with eyes closed, and lying down. If you suspect a chicken is suffering heat exhaustion, get it inside where it's cool and dunk the chicken's feet into a pan of cool water to bring its body temperature down. You don't want to submerge the whole body, just the feet. Holding a wet facecloth to the comb can help in breeds with large combs, since the comb acts as a radiator, expelling heat from the body. Then offer a stronger mixture of 2 teaspoons electrolyte powder per 1 cup of water. Offer to affected adult chickens for several hours as their sole water source, and then offer plain water for several hours. Repeat until symptoms subside. Use only when needed, in cases of emergency.

This electrolyte mixture can also be offered to baby chicks, especially those shipped through the mail. It will give them an added boost of energy.

Discard any unused liquid solution at the end of each day. Dried mixture can be stored in a cool, dry place indefinitely.

A WORD ON NATURAL WORMERS

Chickens can contract internal parasites from wild birds or rodents, from ingesting infected earthworms or slugs, or simply from encountering parasites as they scratch around in the dirt. In fact, many flocks have worms inside them; a healthy chicken will expel the worms on a regular basis, keeping the population in check. It's when the bird is struggling with something else, or not in tip-top condition, that the parasites can get out of control and overrun the system. That's when it becomes a problem. Internal parasites can be detected by a visual poop check (yup, exactly what it sounds like!) or by having your vet test a fecal sample.

Some say that twice-yearly worming is necessary. I disagree. Keeping your chickens healthy and boosting their immune systems with things such as apple cider vinegar and garlic allows your flock to control not only the parasites but also handle lots of health issues that come their way.

Additionally, there are a few foods that are thought to work as natural wormers—and in fact have been used by farmers and homesteaders for decades as preventives. I make it a point to add these foods to my chickens' diet year-round:

» The seeds of fruits in the Cucurbitaceae family, such as cantaloupe, cucumber, pumpkin, squash, and watermelon, are purported to have a coating that works to paralyze the worms.

» Garlic, carrots, nasturtium, and dandelion greens are also thought to help combat worms and other parasites.

» In addition, a bit of blackstrap molasses or plain yogurt can help flush worms out.

HERBAL EGG WASH

As a general rule, I don't wash our eggs until just before using them. (See page 136 for more about that.) However, there is that random dirty egg from time to time—say, if it's been raining and the chickens have mud on their feet. If you've got a dirty egg, or if you're one of those people who just can't bear to keep unwashed eggs in the kitchen, this is for you! I have developed an antibacterial egg rinse to wash any eggs caked with mud or manure, followed by a natural preservative that can be safely applied to replace the egg's natural coating, called the "bloom" or "cuticle," that has been washed off and preserve freshness. This egg wash will easily clean a dozen eggs.

WHAT YOU'LL NEED

Handful of fresh rose petals, crushed by hand to release the oils

Several sprigs of yarrow

Cinnamon stick

Pint-size Mason jar

1 cup white vinegar

1 cup hot water

Spray bottle

Clean kitchen towel

1–2 tablespoons coconut oil

Add the rose petals, yarrow, and cinnamon stick to the Mason jar. These should fill the jar to about the halfway point. Fill the jar with the white vinegar, leaving about ¼ inch headspace at the top.

Let the herbs infuse into the vinegar by leaving the jar on the counter for a week or so, shaking several times to mix the contents. When you are ready to wash your eggs, strain your infused vinegar into a heat-safe container, then add the hot water. You want the mixture to be warm so it doesn't pull any bacteria on the egg inside the shell.

Transfer the mixture to a spray bottle and spray it onto the eggs you need to clean. Use a coarse, clean kitchen towel or your fingers to remove any debris. Dry each egg with a paper towel or let the eggs air-dry. Then gently wipe each egg with a bit of coconut oil or heat the coconut oil and dip each egg into the oil. Set the eggs in a carton to dry. Since the eggs have been coated with coconut oil, they don't need to be refrigerated, but they will last longer in the refrigerator.

The egg wash would make a nice gift for friends or neighbors, whether or not they have chickens. It smells nice and makes a great kitchen or bathroom spray cleaner as well.

HERBAL SINUS TEA

I give my chickens lots of fresh and dried herbs year-round to keep them healthy, but I especially like this recipe for an herbal tea in case you suspect the need for immediate respiratory relief. Offer it alongside regular water to your flock if you hear any sniffles, sneezes, wheezes, or coughs. I use equal parts of each herb, about a cup of each, and then cover the herbs with water in a small pot. Bring the water to a boil, then reduce the heat and simmer for about 20 minutes. Let cool, then serve the tea. You can strain out the solids or leave them for your chickens to nibble on.

» **Sage**—This antiviral and antimicrobial herb helps promote the healthy flow of mucus. Sage is also an anti-inflammatory that relieves sinus congestion.

» **Thyme**—Antiviral and antimicrobial, thyme helps promote the healthy flow of mucus. Thyme also helps warm the sinuses and expel mucus.

» **Calendula**—Calendula is antiviral and antimicrobial, and it boosts the immune system. Calendula is also an anti-inflammatory that soothes the lungs and reduces swelling in the airways.

HERBAL POULTRY POWDER

Providing a dust-bath area for your chickens will typically be enough to keep them free of external parasites, but there are times you might find yourself battling mites or other nasty creepy-crawlies. Dusting your chickens with food-grade diatomaceous earth (DE) or agricultural lime is about the safest treatment you can use that's effective. Nontoxic to mammals and birds, the microscopic shards within DE pierce the shells of hard-shelled insects, eventually dehydrating and killing them.

You can make plain DE even more effective by adding some dried, powdered herbs. Studies done by Newcastle University in the UK in 2006 found that the herbs thyme and lavender work as an effective mite treatment. Try grinding some dried thyme and lavender in a food processor or coffee grinder and then mixing the ground herbs into your DE for added parasite prevention.

Dusting your chickens is easier if you wait until after dark, when they're settled on the roost (this is a great time to check for parasites as well, using a small flashlight). Just apply the DE–herbal mixture under wings and around vents using a plastic squeeze bottle, like the kind diners use for ketchup and mustard. You can also save those plastic Parmesan cheese containers with fliptops. Some of them fit perfectly on a regular Mason jar, turning it into a DE shaker. Use this to sprinkle DE around the feeders to help repel flies.

HERBAL MITE SPRAY

Mites are tiny external parasites that can become a big issue for your flock if they are given the opportunity. Mites are spread by bringing infected birds into a flock, by wild birds and rodents, or by carrying them in on your shoes or clothing. They are more prevalent and active in warm weather and during the summer, although some types do live in cold climates as well.

Mites bite and suck the blood of the host, causing discomfort, feather loss, anemia, or even death in extreme cases. If you believe that your chickens have mites (if you notice excessive preening, feather loss, or tiny spots of blood on the roosts), you might have to get out to the coop after dark to see if you can spot any bugs crawling around on your chickens. If you do, this herbal mite spray is the just the thing to battle them.

WHAT YOU'LL NEED

Pint-size Mason jar

Fresh lavender and thyme

5–6 garlic cloves, crushed or sliced

8 ounces water

Cheesecloth

Lid ring

Fill the jar to halfway with lavender and thyme. Add the garlic and fill to the top with water. Stir to combine. Cover with cheesecloth cut to size and screw on the lid ring over the cheesecloth. Let the herbs infuse for 2 to 3 weeks in a sunny location, shaking the jar several times a day.

After a few weeks, strain the liquid into a spray bottle and spray your chickens every other week as a preventative. You can also use this spray every other day for 2 to 3 weeks in case of an infestation. Concentrate the spray around the vent and under the wings, where mites like to congregate. Spraying the roosts and nesting boxes with the spray is recommended as well, as is changing out all the coop bedding.

COCONUT OIL FOR SCALY LEG MITES

In addition to the various types of poultry mites, which tend to congregate around a chicken's vent or under the wings, scaly leg mites burrow up under the scales of a chicken's feet and legs. The scales on the legs of healthy chickens are shiny and smooth and should lie flat. If you notice the scales on your chicken's legs starting to peel up, flake, or look rough and uneven, it could be suffering from scaly leg mites. If left untreated and allowed to fester, leg mites can affect the chicken's overall health, making it susceptible to other health issues. In extreme cases, leg mites can cause a chicken to go lame or to be unable to roost. Feathered-leg breeds are more susceptible to leg mites.

Traditional treatments have included dipping the chicken's legs in turpentine, gasoline, or kerosene. As you can imagine, none of those is my preferred choice! Instead, try this natural method:

1. Soak the legs and feet in warm water with some Epsom salts, then rinse well and dry them off.

2. Using an old (but clean) toothbrush, gently scrub the legs and feet with white vinegar.

3. Smear the legs and feet liberally with coconut oil (or melt the coconut oil, let it cool, then dip the feet and legs).

4. Keep applying the coconut oil daily until the old scales fall off and new, shiny, smooth scales regrow. This can take several weeks for mild cases and several months for severe cases.

It's prudent to clean your coop at the same time you're treating your chickens. Although the mites live on the host, they can burrow down into the bedding. After removing all the old litter, spray your coop with coop cleaner spray (see page 55), then generously sprinkle food-grade diatomaceous earth on the floor and in the nesting boxes.

APPLE CIDER VINEGAR SOAK

Got a chicken limping or suffering from a foot infection? Spend 10 to 15 minutes soaking your chicken's legs and feet in apple cider vinegar (raw vinegar with the mother is best). In a pinch, it's a great way to fight bacteria and infections; soothe dry, cracked skin; and even relieve minor joint pain. It's thought that the body can absorb some of the goodness in the vinegar through the skin to help balance the body's pH levels as well.

Quick Hack

Steep a chamomile tea bag in boiling water and then let it cool completely. Use the tea bag as a ready-made compress! Press gently to an infected eye for several minutes. Repeat with fresh tea bags several times a day.

CHAMOMILE FOR EYE INFECTIONS

Chamomile is an herb that I've found helpful in treating eye infections. An anti-inflammatory and antimicrobial, it can help heal a chicken with a puffy, infected, or inflamed eye. Brew chamomile tea by

pouring 8 ounces of boiling water over about ¼ cup of fresh chamomile flowers. Let it cool to room temperature, then dip a cotton ball in the tea and drip a few drops into the eye several times a day until the infection clears up. (Use fresh cotton balls for each application, and keep the tea in an airtight container between uses.)

GARLIC FOR IMMUNE-SYSTEM HEALTH

Garlic is one of the cornerstones of natural chicken keeping. Adding crushed garlic cloves to your chickens' water (2 to 3 cloves per gallon) or feeding fresh minced garlic free-choice a few times a week provides them with an immune-system boost, aids in respiratory health, and is thought to help repel external parasites such as mites, lice, and ticks. Garlic is an antiviral, antibacterial, and antifungal, but to get the maximum benefits, the clove has to be crushed or sliced and then exposed to the air for several minutes to release the active ingredient, allicin, before being added to the water.

I will crush or slice the cloves in the house before I head to the coop, and by the time I get down there, have refilled the waterer, and am ready to add the garlic, it has had enough time to release the allicin.

Garlic is also a natural wormer and will reduce the smell of chicken manure in flocks fed garlic regularly. Contrary to popular belief, eggs laid by chickens that eat garlic won't taste like garlic! I like to use fresh garlic when I can. In fact, I even grow my own garlic, since I use so much of it in my chicken keeping as well as my cooking! (See page 162 for simple growing instructions.) However, you can also add garlic powder to your chickens' daily feed. Available in bulk commercially, it is easy to mix into the feed (in a 2 percent ratio to the feed).

6

HACKS FOR YOUR HOME

The chicken coop and run aren't the only areas where repurposing what you have comes in handy. There are also many ways to bring chicken-related projects and hacks into your home. Eggshells, egg cartons, and even eggs themselves can be turned into all kinds of things you can use around your home—candles, fire starters, and common beauty products. I'll show you some of my favorites. I also have a tip to make your coffee taste better! And fresh eggs are notoriously hard to peel when boiled, but I've got a trick to help with that too—as well as an easy way to tell how old an egg is or if it's been hard-boiled or is still raw.

EGG-CARTON FIRE STARTERS

These pretty fire starters are made from old egg cartons. They are not only eye-catching, they're practical. They can be used in a fireplace, wood stove, or fire pit—even when you go camping! Start saving egg cartons as the holidays approach, and make gifts for loved ones. The materials and instructions below use 1 egg carton to make a dozen fire starters.

WHAT YOU'LL NEED

1 cup pine shavings or sawdust

Cardboard egg carton

12 miniature pinecones

2 feet cotton wick cut into 2-inch lengths, plus more to tie carton shut, if desired

2 cups beeswax pastilles, white or yellow

6 cinnamon sticks, broken in half

12 bay leaves

Fresh pine, rosemary, or thyme sprigs

Divide the shavings into the bottom of each compartment in the carton and add a pinecone to each, then prop 1 piece of the cotton wick in each cup, with 1 end trailing over the edge of the cup.

Melt the beeswax until liquid. You can do this however you normally melt beeswax (such as in a tin can on the wood stove). Or, if you'd like to melt it in a glass measuring cup, hang it over a pot of simmering water and stir it with a disposable utensil such as a chopstick. Pour the melted beeswax over the top of the shavings until it nearly fills each compartment.

Note: The wax dries fast, and you'll want it to be melted as you arrange the pieces in the next step—so you might want to pour the wax into just a few compartments at a time, arrange them, and then fill a few more while you keep the melted wax warm in the cup.

While the wax is still melted, arrange the cinnamon sticks and herbs or pine sprigs in the cups, pressing down the contents to arrange them how you would like as the wax starts to dry. Now you can let the wax cool completely. Once the wax has hardened, close the carton and tie it shut with a piece of twine or cotton wick and decorate it with more pinecones or sprigs, cinnamon sticks, and herbs.

To use, cut apart the compartments, snapping the wax. Nestle a fire starter in the kindling and light the wick.

Quick Hack

It's not pretty and it won't smell as nice, but you can use old cardboard egg cartons stuffed with newspaper or dryer lint as fire starters in your wood stove, fireplace, or fire pit!

BEESWAX CANDLES IN EGGSHELLS

Don't have a fireplace or wood stove? You can still enjoy some nice flickering ambience in your home when you make these simple candles using eggshells. If you wish, you can add customized scent to them using your favorite essential oils.

WHAT YOU'LL NEED

1 dozen eggs

Egg carton

2 cups beeswax pastilles, white or yellow

3 feet cotton wick, cut into 3-inch lengths

Essential oils in whatever scent you prefer (lemon, rose geranium, lavender, clove, sweet orange, and wintergreen are some of my favorites)

Butter knife

Measuring cup

Using a butter knife, carefully crack each egg at the narrow end, removing the top third of the shell. Pour the egg whites and yolks into a bowl and refrigerate to use later in a recipe or scramble up for breakfast!

Remove the membrane from the larger bottom part of each eggshell and rinse well, then set upside down on a paper towel to dry. Save the smaller ends you cracked off, air dry, then crush up for your chickens. They'll benefit from a bit of added calcium.

Once the shells are completely dry, set them in the egg carton. Melt the beeswax in a glass measuring cup hung over the side of a pot of simmering water, stirring with a wooden chopstick, or in a tin can on top of your wood stove.

When the beeswax is melted, drop 1 or 2 drops from the chopstick into the bottom of each eggshell and quickly press the wick in to hold it in place until the wax sets. Add a few drops of essential oils, then pour melted wax into each shell using a measuring cup, filling the shells nearly to the top. Make sure the wicks are centered and let the wax harden.

When the wax is dry, trim each wick to the height you'd like. Set the eggs in a ceramic holder, an egg carton, or individual egg cups to use as a pretty table display.

SWEDISH EGG COFFEE

Whether you call it cowboy coffee, campfire coffee, Lutheran church coffee, or Swedish coffee, it's undeniable that adding eggshells will produce a less bitter cup of coffee. Eggshells have the magical ability to pull bitter tannic acid out of the brew. Swedish egg coffee takes it a step further, calling for an entire egg to be added to the coffee pot!

Being of Scandinavian descent, I love this method of brewing coffee. The coffee is brewed without using a filter to allow the coffee's essential oils to remain, which results in a super-rich coffee flavor. The addition of the egg works to separate the coffee grounds from the brewed coffee, while the shell pulls the bitterness and acidity out of the coffee grounds. If you want a smooth, mellow cup of coffee, you need to give this a try.

WHAT YOU'LL NEED

6 cups water

1 fresh egg

¾ cup ground coffee

3 ice cubes

YIELD: 4–6 SERVINGS

Bring the water to a boil in a medium saucepan. Meanwhile, in a small bowl use a fork to stir the egg into the coffee grounds, shell and all. It should look almost like damp potting soil. Add the egg mixture to the water, boil for 1 to 2 minutes, then cover and remove the pot from the heat.

Let stand for 5 minutes, then add the ice cubes and let stand for another minute or so to allow the grounds to sink to the bottom. Ladle the coffee off the top and into mugs. Add the eggy coffee grounds to your compost pile or garden.

Quick Hack

While we're talking about coffee, did you know that you can use eggshells to get the coffee stains out of your coffee mugs? Give this a try: crush an eggshell and put it in your coffee mug. Fill the mug with warm water and let it sit overnight. The next morning, use an old toothbrush to scrub the last of the coffee stains from your mug. Rinse, and you've got a clean mug to fill with steaming hot coffee!

NATURAL EASTER EGG DYES

Dyeing eggs at Easter is a time-honored tradition in many cultures. You can make it even more natural and creative when you use foods and spices from your pantry to color the eggs. You'll soon see there's no need for commercial dyes or food coloring. It's healthier and more economical to use vegetable peels and skins, especially if you grow your own veggies. Although you can use light-brown eggs for this project, you will get brighter colors using white eggs.

WHAT YOU'LL NEED FOR EACH COLOR BATCH

3 white eggs

4 cups water

Small saucepan

Vegetable or spice of your choice (see Natural Dyes Color Chart opposite)

2 tablespoons white vinegar

Wide-mouth pint-size Mason jar

Half an egg carton or piece of paper towel (for drying eggs)

Coconut oil or olive oil

Natural Dyes Color Chart

Bright pink: 2 cups shredded beets

Pale pink: skins from 2 avocados or 2 cups chopped beet greens

Red: 2 cups red onion skins or 2 cups cranberry juice

Bright orange: 2 cups yellow onion skins

Pale orange: 2 tablespoons paprika

Yellow: 2 tablespoons turmeric

Bright green: 2 cups chopped red cabbage and 2 tablespoons turmeric

Pale green: 2 cups chopped spinach

Bright blue: 4 cups chopped purple cabbage

Purple: 1 cup fresh or frozen blueberries

Light brown: 2 cups strongly brewed black coffee in place of water

Start by steaming the eggs as described on page 142 and let cool completely. While the eggs cool, add 4 cups of water to a small saucepan. (You will need a separate pot of water for each dye color.) Add the dye ingredients of your choice to the pot and bring to a boil. Reduce the heat and simmer until the water changes color. This will take about 20 to 30 minutes, depending on your color. Remove from the heat and let the water cool. If you need to reuse the pot to make your next color, you can pour the contents into a bowl to cool. Once the liquid has cooled, strain out the solids, then add the vinegar.

Place the eggs in the Mason jar, then pour the liquid over the eggs, making sure they are covered completely. Leave the eggs to soak overnight in the fridge or until the shells become the desired color.

You're always ready for Easter when you raise a flock of chickens that lay different-colored eggs! Here are some breeds best known for the following colors:

» Cream—Faverolle, Dorking

» Pink—Australorp, Light Sussex, Lavender Orpington, Barred Rock

» Green—Olive Egger, Easter Egger, Favaucana

» Blue—Araucana, Ameraucana, Cream Legbar, Isbar

» Dark chocolate brown—Marans, Penedesenca, Barnevelder

Note: If you want, you can try boiling your eggs right in the dye water so by the time they're cooked, the dye is ready. Then just strain out the solids, put the eggs into the Mason jars, add the vinegar, and refrigerate overnight!

Carefully remove the eggs from the dye with a spoon and place them in an egg carton or on a piece of paper towel on a cookie sheet (to prevent staining your countertop or table) to air-dry. Once completely dry, gently rub the eggs with coconut oil or olive oil and a soft cloth for a shiny finish.

You can reuse the dye water to do another batch. Just pop 3 more eggs into each Mason jar. This time try leaving the eggs in for a shorter or longer time to get a slightly different result. Removing eggs from the dye at various intervals will result in different shades of intensity.

If you plan on eating the eggs, store them in the refrigerator; they will last for up to a week. Since you used natural dyes, you can give your chickens the eggshells to eat. Your chickens will also appreciate the dye water and solids (except for the avocado, onions, and coffee, which can all be harmful to them).

And if you want some breeds that lay white eggs perfect for dyeing, now might be the time to add some white-egg-laying breeds to your flock, such as Anconas, Andalusians, Campines, Hamburgs, Lakenvelders, Leghorns, Polish, Silkies, or Sicilian Buttercups.

EGG HAIR BOOSTER

Any hair type will benefit from the protein, fat, and lecithin in eggs. Egg whites will add shine and volume to your hair, and the protein keratin helps to strengthen hair. Egg yolks add moisture to dry hair as well. What could be better? Well, we'll throw in a bit of lemon juice. Its vitamin C adds elasticity to hair, and its citric acid serves to strengthen hair strands.

WHAT YOU'LL NEED

3–4 room-temperature eggs (depending on the length of your hair)

Juice from 1 lemon

Note: The lemon juice will lighten blonde hair a bit, so if that's not desirable, you can leave it out and just use the egg.

Whisk the egg and lemon juice in a large bowl. Dip the ends of your hair into the bowl or carefully pour the contents of the bowl over your head over the sink. Work into your hair and leave on for 15 to 20 minutes. Rinse off in the shower with cool water for extra shine (and so you don't cook the egg by accident). Feel free to follow up with conditioner if your hair is dry.

Discard any unused egg mixture. Make fresh and use this hair treatment once or twice a month.

EGG-MEMBRANE BANDAGE

Cut your finger while cooking and don't have an adhesive bandage handy? No worries! Just grab an egg. Believe it or not, the membrane of an egg possesses wound-healing properties and is thought to help with pain relief as well. For minor cuts or lacerations, wrapping the wound with an egg membrane can help keep it bacteria free, can curb bleeding, and can speed healing. The albumin in the membrane contains chondroitin collagen and glucosamine, which all help with skin repair. The membrane will let air in to heal the wound while keeping dirt and bacteria out.

To try this, you'll need to remove an egg membrane and apply it to the wound, wrapping it around your finger like an adhesive bandage. (So if you've already cut yourself, a family member may need to help!) As the membrane dries, it should pull the edges of the cut together. Leave the membrane in place until the wound is healed or replace the membrane daily with a fresh membrane. To do this, soak the dried membrane to soften and loosen it, then remove it and put on a new membrane.

Of course, if you have an egg allergy, you shouldn't try this! And use your common sense: for serious wounds, get professional help.

EGG FACIAL

When you spend a lot of time outside, the sun, air pollutants, and wind can wreak havoc on your skin. Taking special care of your face in particular can help keep you looking younger. Eggs contain all kinds of vitamins and nutrients, including collagen and proteins that are beneficial to your body, whether you eat them or apply them externally! Next time you have extra eggs, don't toss them: give yourself a facial!

For dry skin, the egg yolk will be most beneficial. The lecithin in the yolk will help moisturize your skin. For regular or oily skin, you'll want to take advantage of the egg whites instead. The egg white will help to temporarily tighten your skin, shrink pores, reduce puffiness and skin inflammation, help with wrinkles and fine lines, and maintain the elasticity of your skin. The lemon juice acts as a natural astringent and will brighten your skin, as well as remove excess oil. (If you have dry skin and want to gain the benefits of the egg whites, try subbing out the lemon juice and use honey instead. It will add moisture and prevent your skin from drying out.)

Note: If you are concerned about skin sensitivity, you can test the mask on your inner wrist or small portion of your face first.

WHAT YOU'LL NEED

FOR REGULAR OR OILY SKIN

White of 1 egg at room temperature

Squeeze of fresh lemon juice

FOR DRY SKIN

Yolk of 1 egg at room temperature

1 teaspoon honey

For regular or oily skin, whisk egg white and lemon juice until frothy and light. For dry skin, whisk or stir yolk and honey until completely blended.

To use either mixture: First rinse your face with warm water to cleanse it and open up your pores. Then apply your egg blend to your face using your fingertips or a cotton ball. Leave the mask on for 15 to 20 minutes or until completely dry, then rinse your face gently with cool water to close your pores again.

Discard any leftover egg mixture that you don't need. Try this once a week for firmer, younger-looking, healthier skin.

STORING EGGS

The United States is one of the few countries in the world that requires commercially sold eggs to be washed and refrigerated—and that's only been for the last few decades. However, chickens have been laying eggs since well before refrigeration existed, and eggs were stored at room temperature all that time.

How is that safe? Well, when an egg is laid, the last step in the laying process is the application of an invisible coating on the outer shell called the "bloom." This coating works to keep air and bacteria from seeping into the egg through the pores in the eggshell. As the egg ages, air slowly begins to enter the egg, sometimes carrying with it bad bacteria—but an egg will still keep at room temperature for several weeks.

Whether you refrigerate them or not, eggs shouldn't be washed until you're ready to crack them, and they should always be rinsed under warm running water (at least 20 degrees Fahrenheit warmer than the shell temperature to prevent any bacteria on the shell from being pulled inside the egg). Never let uncooked eggs sit in a bowl of water, as this can transfer bacteria.

Quick Hack

The cleaner you keep your nesting area, the less chance you will be collecting dirty eggs. You want to discourage your chickens from sleeping in their nesting boxes (as discussed on page 47, blocking the boxes at night can help with that). Refreshing (or at least checking) the bedding each morning when you open up your coop and replacing the bedding if necessary ensures nice clean nests for that day's eggs. And the more often you collect the eggs, the less chance they'll get broken or end up dirty.

DIY WOODEN EGG HOLDER

Since fresh, unwashed eggs don't need to be refrigerated, feel free to display your prettiest eggs right out on the kitchen counter. This egg holder is functional as well as attractive. It's a great way to repurpose an old cutting board that has seen better days. If you don't have a cutting board you're willing to cut holes in, they're pretty easy to find inexpensively at secondhand shops or you can use a board cut to size.

WHAT YOU'LL NEED

6½" x 9½" or similar-sized cutting board or scrap piece of lumber

Cordless drill and 1½-inch drill bit

Beeswax oil finish (cutting-board wax)

4 wood screws slightly longer than your board is thick

Using a pencil, trace around the drill bit where you will be making your holes in a 3×4 pattern, leaving at least ½-inch between holes and a 1-inch border around the edges. Using the drill bit, drill holes through the board where you have them marked. (Save four of your hole cutouts to make some cute legs for your holder!)

Once your holes are finished, lightly sand any rough edges, and then use the screws to screw the cutouts onto the underside of the board—1 at each corner—to make short "legs" for your holder.

Buff all the exposed wood with food-safe cutting-board wax. Then fill the holder with your prettiest eggs!

FREEZE EGGS FOR THE WINTER

Egg production naturally slows as the days get shorter in the fall. Yet there's always more baking rather than less as you get into winter! That's one reason I like to freeze some eggs while production is still high at the end of the summer. That way I have delicious, fresh eggs to use for Christmas cookies and cakes. Below, you'll find my tips for freezing eggs a few different ways.

Freezing whole eggs: To freeze whole eggs for baking or scrambling, lightly whisk the eggs with a fork to combine the yolks and whites, and add a pinch of salt. The salt is optional, but if you skip it, your eggs may have a grainy texture. Pour the mixture into ice-cube trays and freeze them. Silicone trays work well and make the eggs easier to get out once they're frozen, although spraying your ice-cube tray with coconut oil or other cooking spray is helpful as well. Once the cubes are frozen, you can pop them out and store them in freezer bags. Defrost the cubes as needed overnight in the refrigerator or on the counter. Each 3-tablespoon cube is the equivalent of 1 egg, so it's very easy to measure the out the correct number of cubes for recipes.

Freezing egg whites: Some recipes, such as meringues, call just for the whites of eggs. To freeze

egg whites, separate the egg and drop each white into a compartment of the ice-cube tray. Freeze, store, and defrost as for whole eggs.

Freezing egg yolks: Other recipes, such as Hollandaise sauce or lemon curd, call for just the yolks of eggs. To freeze egg yolks separately, you will need to separate your eggs and collect the yolks in a bowl. Lightly whisk with a fork, then add a pinch of salt. Measure out the yolk mixture with a tablespoon into the ice cube tray. One tablespoon of yolk is the equivalent of 1 egg yolk. Freeze, store, and defrost as above.

You can also freeze whole eggs if you want to make fried eggs or egg sandwiches. Just break each egg into an ice cube tray. Freeze, store, and defrost as above.

Be sure to use frozen eggs immediately once they're defrosted and only in recipes that call for the eggs to be fully cooked. Frozen eggs will last for about 6 months in the freezer.

Quick Hack

If you want to quickly separate eggs, try this: crack an egg into a small bowl, being careful not to break the yolk. Take a plastic water bottle and hold it just above the yolk, squeeze the bottle, release, and suck the yolk inside the bottle. Then squeeze the bottle again to drop the yolk into the frying pan, ice-cube tray, or mixing bowl.

STEAM AND PEEL EGGS

The one downside to having eggs fresh from your coop is that they are impossible to peel! As an egg ages and air gets in through the pores, that air forms a pocket between the membrane and the shell, making it easier to remove the shell once the egg is cooked. You can let your eggs sit around and get old if you want to make deviled eggs or hard-boiled eggs. Or, you can try this brilliant trick for perfectly peeled *fresh* eggs: steam them!

Here's what you do: Heat water to boiling in a pot. In the meantime, rinse your eggs under warm running water to clean them. Then set them in a single layer in a colander, double boiler, vegetable steamer, or bamboo steamer over the pot of boiling water. Cover the eggs and steam them for 20 minutes. Remove the eggs and let them cool in a bowl of ice water. Once cool, peel them and watch the shells come off like a charm.

When you're ready to peel, crack an egg against a *flat* surface to break the shell. This is to prevent shards of shell (and possibly bacteria) from getting into the egg white or yolk, which can happen more easily if you crack the shell against the sharp edge of a counter or bowl.

Can't remember if the bowl of eggs in the fridge are raw or hard-boiled? Try this quick trick. Place the egg on a flat surface, such as the kitchen counter or a cutting board. Give the egg a quick spin. A hard-boiled egg will spin smoothly, while a raw egg will wobble. And if you put a finger on the egg to stop it from spinning, the cooked egg will stop immediately while the raw egg will continue to spin for a bit. As a last resort, shake the egg. If you feel sloshing inside, the egg is raw.

Quick Hack

Next time you have the grill going, put a few eggs on while you're grilling your steaks or burgers. Let them cook for about 20 minutes, then plunge them into a bowl of ice water until cool. You'll have hard-cooked eggs with a slightly smoky flavor for the next morning's breakfast!

THE FLOAT TEST

While eggs are perfectly good to eat for far longer than you think—several months if handled properly—they will go bad at some point. And there's nothing worse than a rotten egg! The more you can do to preserve the integrity of the bloom on an egg (the invisible coating on the shell that prevents air and bacteria from entering the egg), the longer the egg will last.

Unwashed eggs will last for several weeks out on the counter and even longer in the refrigerator. (Eggs will last 7 times longer if they're kept in the fridge.) However, eggs do eventually go bad. A rotten egg will be visibly bad (you'll start to see a black shadow through the shell, or the egg might start weeping), and your nose will smell it from a mile away. If you're not seeing or smelling rottenness but you're still suspicious, here's a quick trick to figure out just how old those eggs you found in the back of the fridge really are!

Quick Hack

Write the date you collect an egg in pencil on the shell to keep track of which eggs are oldest.

Gently drop the egg into a glass of water. If an egg sits flat on the bottom of the glass, it's newly laid. As an egg ages and air gets into it through the pores in the eggshell, the egg will start to float and eventually will rise up off the bottom of the glass. As air seeps into the egg, one end will start to lift. The egg is still pretty fresh, but it has begun to lose some of its nutrients. As the egg continues to age and more air enters through the shell, the egg will start to sit at a steeper angle. These eggs are still perfectly good to eat (and likely still fresher than your average supermarket egg). Eventually the egg will be standing almost straight up. The quality of an egg at this stage has been degraded. There's a chance bacteria has entered the egg as well, but as long as one end is still sitting on the bottom of the glass, believe it or not, the egg is likely still good to eat (and great for hard-boiling, since it will peel easily).

On the other hand, if an egg floats in the glass, it signals that egg is very old and most likely has gone bad. I would throw it out. It *might* still be good to eat, but why take a chance when so much air (and possibly bad bacteria) has had a chance to penetrate the shell? In any case, if you're unsure of the age of an egg, it's good practice to crack it into a small bowl first to be sure it looks and smells okay. You wouldn't want to ruin your whole recipe by cracking a spoiled egg directly into it.

MASON-JAR MAYO

One of the best things about having fresh eggs is that you can have more peace of mind when using those eggs in recipes where the egg isn't completely cooked. The risk of salmonella should be far less than store-bought eggs, especially when you only use your cleanest, freshest eggs for any recipe calling for uncooked eggs. One of my favorite things to make with my chickens' eggs is mayonnaise. Once you've tried this easy 1-jar recipe, I'm sure you'll agree. It's that delicious and that simple to make.

WHAT YOU'LL NEED

3 egg yolks

Note: This recipe uses raw eggs. Very young people, very old people, or anyone pregnant, nursing, or with a compromised immune system should use caution when eating raw eggs, due to the slight risk of salmonella.

½ teaspoon coarse stone-ground mustard

Juice from half a lemon

¼ teaspoon salt

Wide-mouth pint-size Mason jar

Immersion blender

1 cup cooking oil (any type)

Measuring spoons

YIELD: ABOUT 1 CUP

Add the egg yolks, mustard, lemon juice, and salt to the Mason jar. Using the immersion blender, pulse a few times to combine the ingredients. (You can also use a regular blender.) Once the ingredients are combined, start adding your oil slowly, 1 tablespoon at a time, pulsing until the oil incorporates into the egg-yolk mixture. Once you've added about half the oil, start to slowly drizzle the remaining oil into the Mason jar, continuing to pulse it with the blender.

Your mayonnaise should start to thicken and lighten to a lemon-yellow color. Continue to blend until it turns into a nice, spreadable consistency. Use your mayonnaise immediately or put the lid on the Mason jar and leave it on the counter at room temperature for up to 12 hours, then refrigerate it. (This allows the acid in the lemon juice to help to kill any bacteria.) Your mayonnaise will last up to a week refrigerated.

Want to make flavored mayo? Try one of these combinations and finish to taste with salt and pepper:

» **Garlic**—Stir in a minced garlic clove.

» **Herb**—Stir in ¼ cup of minced fresh herbs (or about 1 tablespoon dried). Some favorites of mine are basil, tarragon, and dill.

» **Tartar sauce**—Stir in 1 tablespoon of chopped pickles or relish, a teaspoon of chopped onion, and a pinch of dried dill.

» **Maple bacon**—Stir in 2 tablespoons of crumbled cooked bacon and 1 teaspoon of maple syrup.

» **Pesto**—Stir in 1 tablespoon of fresh chopped basil, 1 minced garlic clove, 1 teaspoon of grated Parmesan cheese, and 1 teaspoon of ground walnuts or pine nuts.

» **Lemon caper**—For this personal favorite of mine, stir in 1 tablespoon of chopped capers, 1 tablespoon fresh chopped parsley, a minced clove of garlic, and a bit of additional lemon juice.

CHAPTER 7

IN THE GARDEN

Gardening and chicken keeping naturally complement each other. If you do one without the other, you're missing out on the synergy that results from using some of the natural byproducts of chicken keeping in your garden. I also believe in having your chickens help out in the garden with some natural pest and weed control. Furthermore, by growing your own food, you're automatically saving money on your grocery bill—not to mention eating healthier! In this chapter, we'll look at using not only chicken manure (the *best* garden fertilizer and for us, as it's free!) but also eggshells and feathers in the garden. In addition, I'll share ways to reuse your egg cartons and feed bags. If you want to dig deeper into ways your chickens can be put to work to help out in the garden, as well as crops to grow for them and more about the benefits of chicken manure, check out my book on the topic: *Gardening with Chickens*.

CHICKEN-POOP TEA

Chicken manure is one of the best garden fertilizers out there. However, with its superhigh nitrogen content, it can burn plants if it's applied directly. There's also the pathogen issue to worry about; chicken manure, like other types of feces, can contain bacteria such as salmonella or *E. coli*. That means it's best to let your chicken manure age before using it on the garden, either by applying it to your soil in the fall so it ages through the winter and is ready to incorporate into the soil come spring, or by composting it. Another fun way to use your chicken manure is to brew it into chicken-poop tea, which is the ultimate liquid fertilizer. For the cost of a plastic bucket, you can make your own garden fertilizer for free! By steeping the manure for a few weeks, you allow all the great nutrients, good bacteria, microorganisms, and enzymes to leach into the water, providing a healthy drink for your plants. You'll want to do this in the warm months, because the manure needs to age outside for a couple of weeks, and you don't want it to freeze! Early spring is perfect, so you can offer it to your seedlings.

WHAT YOU'LL NEED

Old pillowcase

Fresh chicken manure (enough to fill about ⅓ of the pillowcase)

Piece of twine or ribbon

5-gallon bucket or plastic pail

Watering can

When you scrape off your dropping board or rake out the run, collect a bucketful of the manure. Fill the pillowcase about ⅓ full with manure. Tie the top of the pillowcase shut with the twine or a pretty ribbon and then set it in the pail with the ends of the twine hanging over the edge. Fill the pail with water. Be sure the contents of the pillowcase are completely submerged. Set the pail outside (as long as the outside temperatures are staying above freezing—and closer to 65–75 degrees Fahrenheit is optimal) where it won't be disturbed (and the smell won't bother you). Agitate the water a few times a day to keep the liquid oxygenated.

After 1 to 2 weeks, the liquid should be a nice brown tea color. Remove the pillowcase from the pail and empty the solid contents back into your compost pile. Dilute the remaining tea, 1 part tea to 4 parts water. You can use a watering can to apply the chicken-poop tea to the soil around the base of your seedlings every other week. Store leftover tea outside in the original pail.

Manure tea is most useful when given to young seedlings and plants for a boost of nitrogen to help them grow nice green leaves. You can continue to water them until they start to flower, then reserve the manure tea for plants that seem stressed or that need a bit of a boost. Use care, though: too much nitrogen applied to mature plants can result in an abundance of leaves but stunted fruit growth.

Note: Because of the potential for pathogens, always wash your hands when brewing or applying the tea, don't apply it to root crops (beets, carrots, potatoes, radishes, and so on), and don't apply it directly to leafy greens you will be consuming (lettuce, kale, spinach, and so on). Watering around the base of your plants is the best way to apply manure tea. Giving the chicken manure time to age (3–6 months) will reduce the risk of pathogens and also allow the nitrogen in the manure time to dissipate, making your tea less "hot" and still extremely beneficial but less likely to burn your plants or transmit any disease.

TURN OLD WATERERS AND FEEDERS INTO PLANTERS

Don't throw out your old, dented, rusty metal feeders and waterers. While your chickens shouldn't drink from a rusted waterer or eat from a rusty feeder, you can turn these items into beautiful planters for flowers or herbs. Simply put some small stones, pieces of bark, pinecones, straw, or other material that will help with drainage in the bottom, then fill with potting soil. Plant seeds or seedlings, water well, and set in partial sun. Since the metal will warm up very quickly in the sun, be sure to move your planter into the shade during the hottest part of the day or the warmest time of year if the plants seem to be suffering. Keep it well watered; the soil should be moist, but not waterlogged. Herbs like sunny conditions and fairly dry and sandy soil, so they're a good choice for your new planters.

Quick Hack

If you don't want to use your vintage feeder, knock a piece out of the rim of a terra cotta flowerpot and set it upside down in your garden for the toads.

VINTAGE FEEDER TOAD HOUSE

If you're lucky enough to have a vintage stoneware poultry feeder top passed down from your grandparents, this is the hack for you! (If you don't, you can often find old feeders on eBay or at antique shops fairly inexpensively.) Why make a toad house? Toads are amazing hunters and will help control flies and other insects around your coop. Setting the toad house under a bush by your coop can attract toads looking for a nice, cool, safe spot to spend their day. Just scoop out a bit of dirt underneath the feeder and in front of the "doorway," then set a shallow dish of water out for any toads that might wander by and decide to stay.

RAID THE COOP FOR GARDEN MULCH IN THE FALL

Each fall when I clean out my chicken coop for the winter, I rake all the old straw, feathers, and chicken manure into a wheelbarrow and then spread the mixture over my garden, concentrating on the area where I have planted my garlic. This way the manure will have several months to age and decrease the nitrogen levels a bit, while it acts as an insulating blanket over the garden soil. It will keep the temperature more consistent and retain more moisture through the winter while also working to control weeds.

any seedlings I thin out. Before raising chickens, I always dreaded the job of thinning seedlings, because I felt like it was such a waste to plant seeds, then pluck out half of the seedlings to leave room for the remaining plants to grow. Of course I composted the thinned seedlings, so they didn't go completely to waste . . . but now that we have chickens, I look at it as providing supernutritious, yummy treats for them, which makes me (and them) really happy!

Your chickens will also love carrot tops, radish and beet greens, peas, and beans. Just avoid sharing the leaves, stems, or unripe fruit from plants in the nightshade family, which includes tomatoes, peppers, white potatoes, and eggplant, as this plant family contains toxins that can be harmful to chickens. Rhubarb leaves are also toxic to chickens, as are members of the onion family.

USE EGGS TO KEEP DEER OUT

Deer don't like the smell of eggs. They can purportedly smell eggs long after they are no longer discernible to the human nose. Spraying this natural deer repellent on your garden plants can help protect them from being ravaged by deer.

Combine 3 eggs, beaten, with 1 cup of milk and close to 1 gallon of water in a gallon milk or water jug. Let sit for a week or so, uncovered, then pour into a spray bottle and spray your garden plants. You should only need to reapply it about once every 1 to 2 weeks, or after a heavy rain.

CHICKEN FEATHERS AS FERTILIZER

Chicken feathers are made up mostly of keratin, which is a protein. As the feathers fall out, say while the chicken is preening or during the molt, the keratin eventually breaks down and provides beneficial nitrogen to the soil. Raking the feathers out of your run or coop in the fall and spreading them over your garden is wonderful for your soil. The feathers will take a while to decompose, which means that they will release nitrogen slowly for your garden over the course of several months.

TREATS FROM THE GARDEN

When you're out weeding or doing some bug control in the garden, don't forget about the chickens! I bring a pail out with me into which I toss bugs, insect larvae, wilted or bug-eaten leaves, and

Quick Hack

Still struggling with deer? Planting garlic around the perimeter of your garden can also help deter deer, as can ringing your garden with a single strand of fishing line strung onto garden stakes or posts about 3 feet off the ground.

Quick Hack

Deer aren't the only garden marauders. Wild turkeys and other birds, as well as your chickens, will also wreak havoc on your plants. So don't toss your old bent, rusted tomato cages! Wrap them in chicken wire, and you've got instant armor to keep your plants safe from your chickens or other garden invaders. Tie a short length of clothesline to the top of each cage to use as a handle. They're great for berry bushes, rosebushes, peonies, lilacs, or any other tree or shrub while it's still small and the leaves are at chicken level.

PICTURE-FRAME
Herb Drying Rack

Herbs benefit from frequent plucking, snipping, and trimming, and they're really easy to air-dry. That's why I love this homemade herb drying rack made out of old picture frames. I prune my herbs all throughout the growing season and dry the herbs in batches, both for cooking and to add to the chickens' feed and nesting boxes through the winter. You can pick up old wooden picture frames inexpensively in thrift or secondhand shops and at yard sales—or maybe you have some in your garage or basement that you can repurpose.

WHAT YOU'LL NEED

3 wooden picture frames in varying sizes (any size will work, but I like using a larger frame at the bottom and then consecutively smaller frames on top, such as a 16×20, 11×14, and 8×10)

Paint (optional)

Staple gun and staples

Piece of window screen (save those old screens if you ever have to replace them!)

Cordless drill and small drill bit

20 small eyehooks

12 lengths of chain (7–8 inches each)

Metal ring

Remove the artwork and glass from the frames. Paint the frames, if desired, and let dry. You can also leave them as-is or stain them. Once the frames are dry, flip them over and staple a piece of window screen to the back of each. Then arrange the frames in descending size order.

Predrill a hole in each of the 4 corners of each frame. You'll need holes in both the top and bottom of each frame, except the bottom of the lowest (and presumably largest) frame. Screw an eyehook into each hole. Then, using the chain, attach the frames together. Be sure to leave 7 to 8 inches between the frames to allow room to easily access the herbs as they dry. Connect the 4 lengths of chain attached to the top frame to the metal ring. Now your whole rack can hang from a hook, and it will quickly collapse for storage during the off-season.

HARVESTING AND DRYING HERBS

It's best to harvest herbs in the morning, after the dew on them has dried but before the sun is at its peak. With your fingers, or a pair of kitchen shears or small scissors, cut the leaves from the stems. You can prune herbs pretty aggressively, but you never want to take more than ⅓ of any plant.

Lay the herbs in a single layer on the drying rack and hang the rack in an out-of-the-way spot, out of the direct sunlight, where the herbs won't blow away. Since I use my rack exclusively for the chickens, I'm not concerned about hanging this right in my potting shed. Once the herbs are dried,

crush them with your fingers and store them in an airtight, covered container in a cool, dark spot, such as a kitchen cabinet or pantry.

I don't separate all the herbs into separate batches to dry, but I do separate them into those that will be dried, crushed, and added to the chickens' feed and those that will be dried and added to the nesting boxes. This simplifies my drying into just 2 batches. My feed herbs (about 2 cups of dried herbs per bag of feed) include basil, cilantro, dandelion, dill, echinacea, marigold, marjoram, nasturtium, oregano, parsley, sage, tarragon, and thyme.

Quick Hack

Don't feel that ambitious? If you have an old garden rake, just bang some large nails into the wall of your coop and suspend the rake from the tine end. Then hang bunches of cut herbs from the tines and let them air dry.

BODY BUTTER FOR DRY HANDS

Even the cleanest chicken yards can potentially transmit salmonella and other pathogens to humans. Washing your hands after handling your chickens is critical. When you also garden, you'll find you need to wash your hands even more often! All that hand washing can dry out your hands, so whip up a batch of this all-natural, smooth, rich salve.

Not only will the coconut oil and sweet almond oil in the salve moisturize your hands, the citrusy addition of lemon oil will help to cleanse, rejuvenate, and soothe your skin. Lemon oil can also help relieve muscle stiffness and joint pain, and it has antiseptic properties, guarding small cuts from infection. As a nod to our garden helpers, the bees, I've used beeswax as the base and added honey to the mix. A dash of vanilla essential oil softens up the scent.

WHAT YOU'LL NEED

4 ounces beeswax pastilles or grated beeswax

4 ounces coconut oil

4 ounces sweet almond oil

2 tablespoons honey

Hand mixer

20 drops lemon essential oil

20 drops vanilla essential oil

Wide-mouth half-pint Mason jar with lid

Measure the beeswax and coconut oil into a medium-size glass mixing bowl that's been set over a pot of simmering water. Keep the heat at medium–low, just simmering, and melt, stirring until the liquid is clear. Stir in the sweet almond oil and honey and let cool. (You can put the mixture in the refrigerator or freezer to speed things up a bit; it will take a few hours to solidify at room temperature.)

Once the mixture has cooled and turned cloudy, use the hand mixer on low speed to whip the mixture, scraping down the bowl sides as needed. Add the essential oils and then increase the mixer speed to high and continue whipping until your body butter starts to look like whipped cream and stiff peaks form. (This will take several minutes.)

Scrape the butter from the bowl and spoon it into the wide-mouth half-pint Mason jar. To use, scoop out a generous amount of the body butter and massage into your hands (or other dry areas, such as feet, elbows, or calves) as desired.

WINDOWSILL HERB GARDEN

Extend the life of your fresh herbs by bringing them inside in the fall. Don't bother digging them up and repotting them. Make a water herb garden instead! Just before the first frost, or even a bit later for a few of the hardier herbs, trim the tops off your basil, parsley, rosemary, lavender, sage, and other herbs—cutting just above a set of leaves—and put them into small containers of water on your kitchen windowsill. This will extend their life, providing you with fresh herbs to cook with and to share with your chickens, well into the winter. Just be sure to pluck off any leaves below the water line and change the water once a week. You can snip off individual leaves as needed; just be sure not to take more than ⅓ of the plant at any time.

STRAW BALE–WINDOW COLD FRAME

Starting seeds indoors is a great way to get a head start on your gardening. You can get them planted weeks before the last frost and bring them outside for a few hours a day to harden them off before planting them in the ground. Hardening off is best done in a sheltered area, but greenhouses are expensive. How about making a cold frame instead? All you need are 4 straw bales and an old paned window to form an inexpensive cold frame. A cold frame is also a great way to extend the growing season. If you have your herbs planted in pots, for example, you can pop them inside the cold frame at night to keep them from being killed by the first few light frosts.

GROWING GARLIC

Since I use so much garlic in my chicken keeping as well as my cooking, I've found it's much more economical to grow my own. Although you'll need to buy a few bulbs to start with—organic from your farmers' market is best—once you harvest your first crop, you can save a few of the best cloves to replant and have free garlic from then on!

PLANTING GARLIC

Plan on planting your garlic about 5 to 6 weeks before your first anticipated frost in the fall. You want to give the cloves time to sprout some roots before winter sets in, but not enough time to do any meaningful growing. Choose an area that gets full sun and drains well. Due to its pungent aroma, garlic is rarely bothered by insects, moles, rabbits, deer, or even chickens, so don't worry about fencing it in.

Once you have some roots, break the bulb into cloves, leaving the papery covering on them. Plant the cloves pointy side up about 6 inches apart and with the tip 2 inches below the surface. Cover them loosely with dirt and then mulch the cloves with about 4 inches of chopped straw, dried leaves, or hay.

Here's a money-saving hack: I use the straw from the coop when I do my fall clean-out. The mulch will moderate the soil temperature through the winter, which helps the roots remain in place. It will also retain moisture. (Garlic doesn't need to be watered daily unless your area is extremely dry and hot.) A light watering every few weeks is helpful, but too much water can cause the bulbs to rot, so err on the side of less.

In early spring, when the green shoots start to poke through the ground, you can carefully remove any remaining mulch, although I prefer to leave it in place to continue to control weeds and retain moisture. The chicken manure in the old coop

bedding decomposes through the winter, releasing nitrogen and other nutrients into the soil and eliminating the need for additional fertilizer.

In late spring, look for scapes to appear. These are the round, thin, curly stems that grow up from the center of the foliage. The scapes will eventually form flowers, which drain the energy needed to grow the new bulb, so you'll want to snap them off. If you don't remove them, you'll end up with smaller garlic cloves. (Don't discard them! They're delicious cut into short lengths, tossed into a cast-iron skillet, and seared with some olive oil, salt, and pepper.) Continue checking for the scapes and removing them through the late spring and early summer.

HARVESTING GARLIC

The garlic is ready to harvest when the foliage has turned yellowish brown and starts to fall over. Using your fingers or a small rake or trowel, carefully dig up the bulbs. Gently wipe off any dirt from the bulbs and then leave them to dry in an airy, cool spot out of direct sunlight for 2 weeks. If you're ambitious, you can braid the leaves or tie them into bunches, then hang them to dry. Or you can just lay them on a clothes drying rack, oven rack, or sheets of newspaper—or loosely arrange them in a wooden crate or other open container.

After a few weeks, once the wrappers are dry and papery, you can cut the tops and roots off and store your garlic bulbs in the pantry or root cellar. If you braided the bulbs, you can leave them hanging in your kitchen or pantry and cut them off as you need to use them.

SWEET-POTATO GROW BAG

Potatoes are a lot of fun to grow, and despite what you might think, you don't need a ton of space to grow them. They'll be perfectly happy growing in a bag on your patio. Empty feed bags make perfect (free!) containers to grow potatoes in as well! While white potatoes are in the nightshade family, and every part of the plant contains a toxin that isn't good for your chickens, sweet potatoes are safe as they're in the morning-glory family. They are completely edible—leaves, stems, and all—and more nutritious for your family and your chickens. While they are more fragile and intolerant of cold temperatures than white potatoes, it's possible to grow them in northern climates. The advantage of planting them inside a grow bag is that you can start them earlier indoors if you live in a cold climate, you can move them around to keep them in full sun, and they're easy to harvest if you slit the bag open. Ready to try growing your own sweet potatoes? It's fun and easy!

WHAT YOU NEED

Empty feed bag

Small piece of window screen

Potting soil

Straw

Sweet-potato slips (basically sprouted potatoes that you'll grow your new plants from)

Mason jar or small container (if growing slips, see page 165)

Flatten the bottom of the feed bag and cut several holes in the bottom for drainage, about 1 inch in diameter. Cut a piece of window screen to fit in the bottom of the bag. This will keep the soil in the bag. Now roll down the top until your bag is about a foot tall.

PLANTING YOUR POTATOES

Once the danger of frost has passed, fill the bag with about 4 inches of potting soil mixed with some straw. Tuck three or four slips into the soil and then cover with 4 inches more soil/straw mix and water well. You don't want to fertilize your potatoes. That will encourage foliage growth instead of root growth and lead to a smaller harvest. Set the bag in the sun and bring it indoors on cold nights.

As your plants start to grow, roll the top of the bag up little by little and mound more soil around them, keeping all but the leaves buried in the soil. Keep the soil moist, but be sure not to overwater.

Sweet potatoes are cold sensitive, so you should wait until about a month after your last frost to plant them.

That said, in the grow bag in the sun, the soil should be a bit warmer than your regular garden soil. If you live in a northern climate, you can plant your slips in the bag indoors to give them an earlier start and then bring the bag outside once the temperatures warm up. The sweet potatoes need temperatures above 60 degrees or so and prefer soil temperatures between 70 and 80 to thrive. The advantage of the bag is that you can keep it in the full sun and the soil in the bag will stay warm, and you can also move the bag indoors if a cold night is predicted in early spring.

HARVESTING AND STORING YOUR POTATOES

It will take between 4 and 5 months until your potatoes are ready for harvest, depending on the variety you grow and where you live (be sure to choose a short-season variety if you live in zone 4 and 5, for example). When the leaves turn yellow and start to die, your potatoes should be mature. Even if they are not as large as you'd like, you do need to harvest them before the first frost.

To harvest, cut the side of the bag and carefully pull out the potatoes or dump out the bag, if you want to use it again. (Add the dirt to your garden if you'd like.) Brush the dirt off the potatoes, then lay the potatoes out to air-dry for several weeks at room temperature. At this point you can place them in the pantry or in a box, paper bag, or basket in a cool, dark, dry location around about 55–60 degrees for long-term storage. You can also cut the potatoes into chunks and freeze them. Your chickens can eat them raw or cooked, peels and all. You can either till the stems and leaves into your garden or give them to your chickens to enjoy. Be sure to save some potatoes to start next year's crop!

Starting Potato Slips

Sweet potatoes don't grow from an "eye" like a white potato. Instead, you need to start a plant from the potato itself. So save a few potatoes from each crop to plant the following spring. To start the slips, 6–8 weeks before you're ready to plant outside, submerge ⅓ to ½ of a sweet potato in a jar of water on your kitchen windowsill, or lay the potato on its side in a shallow container filled with moist soil in a warm, sunny spot in your house. Keep the soil very wet. Your potatoes should start to sprout. Once the sprouted vines are about 6 inches long (it will take a few weeks), break them off the potato and place them in the jar of water on the windowsill until they grow nice strong roots. When the vines have sprouted roots, you can proceed as described in this project to plant them. Alternatively, you can plant the potato, vines and all, right in your grow bag.

WINDOWSILL EGGSHELL HERB–EDIBLE FLOWER GARDEN

If your herbs are growing too fast and producing more than you can use, why not make a pretty windowsill arrangement? Crack the tops off 6 eggs, pour the contents into a bowl, and rinse the shells. (Save the egg whites and yolks for a future recipe.) Set each shell into an egg carton or ceramic egg tray, fill with water, then arrange sprigs of herbs and edible flowers in each shell. Trim and use the herbs and flowers as needed in your cooking.

CRUSHED-EGGSHELL CALCIUM BOOST

Some vegetable crops, such as tomatoes and peppers, really appreciate an extra dose of calcium to prevent a condition called blossom end rot. Your extra eggshells, crushed and scattered around the base of those calcium-loving plants, will provide the extra calcium they need. As an added bonus, eggshells can help keep slugs from damaging your crops, such as lettuce, cabbage, and other tender leaves and shoots.

STARTING SEEDS IN EGG CARTONS OR EGGSHELLS

Come spring, start saving your cardboard egg cartons and eggshells. Both make for nice, inexpensive seed-starter cups. For an egg carton, fill each compartment with potting soil, add your seeds, and water. Set the egg carton on a sunny windowsill and keep the soil moist. When you're ready to plant the seedlings, carefully cut the sections apart and plant them right in the ground. The cardboard egg cartons will disintegrate into the soil as they decompose.

Eggshells make a great natural pot that will slowly release calcium into the soil for the seedlings as they grow, which is especially beneficial for calcium-loving plants such as tomatoes, eggplants, and peppers. Carefully crack the top off the shell, pour out the contents to use later in a favorite egg recipe, and poke a small drainage hole in the bottom of the shell with a pin or sewing needle. Fill each shell with potting soil, and add the seeds. Water well using a spray bottle, then set each shell in an egg carton and put on a sunny windowsill. Water as needed to keep the soil moist. Once the seedlings are ready to be transplanted into the garden, crush each eggshell a bit with your fingers and dig a hole for it in the ground. Plant it, shell and all.

Quick Hack

Egg cartons shouldn't be reused for eggs you're selling or giving away, due to biosecurity reasons and to prevent the spread of germs and bacteria that might be on the eggs previously in the carton. However, they make wonderful compost material for your compost pile. So don't toss them, recycle them in the garden!

CHICKEN TUTUS

I don't use, nor do I recommend, chicken sweaters or jumpers in the winter. Such heavy clothing impedes chickens' ability to fluff their feathers to stay warm. Nor do I advocate chicken diapers. (Chickens don't belong in the house; they're still livestock, much as we love to name them and treat them like pets.) Still, I have to say that there's nothing cuter than a flock of backyard chickens running around in tutus!

They're perfect for those Instagram photo shoots, a kids' birthday party, Halloween, or just pure entertainment, and the chickens really don't seem to mind them. Dare I say my chickens actually enjoy pirouetting around the yard in them? It may go without saying, but tutus are for entertainment purposes only. They should be used only under adult supervision and removed when the party is over (or if the chickens show any signs of distress)!

And while this has certainly not been scientifically studied nor proven, I would venture a guess that a flock of chickens pirouetting around the yard in colorful tutus would give your average fox or hawk reason to pause and wonder exactly is going on!

WHAT YOU'LL NEED

Piece of elastic

Rolls of tulle in various colors

Scissors

Measuring tape

Measure around your hen under her wings and in front of her legs, and add 1 inch to that measurement. Cut the piece of elastic to that measurement and knot the ends to form a circle.

Cut the tulle into 25 pieces, each measuring 12×3 inches. Fold each length of tulle in half the long way, then loop it over the elastic and pull the ends through the loop to create a knot. Continue around the elastic until your tutu is as full as you want it to be. Carefully slip it over your hen's head and gently pull her wings through the tutu so it rests comfortably under her wings.

About the Author

Lisa Steele is a fifth-generation chicken keeper who has been around chickens most of her life. It's no surprise that she has made her name raising her own backyard flock and sharing her farming adventures on her popular website, Fresh Eggs Daily. As an aspiring herbalist and avid gardener as well as a thrifty New Englander born and bred, she also enjoys mixing up herbal concoctions and tackling DIY projects for the coop and run.

A regular contributor to such publications as *Backyard Poultry*, *Hobby Farms*, *Old Farmers Almanac*, and *Chickens* magazine, as well as HGTVGardens.com and BHG.com, Lisa has become the most trusted voice in natural chicken keeping. She regularly appears on national and local television as well as radio shows, offering advice and encouraging listeners to dive into backyard chicken keeping.

Fresh Eggs Daily has been named one of *Better Homes & Gardens'* Top Ten Gardening Blogs, and her Facebook page has attracted over half a million followers who enjoy her down-to-earth, practical advice. She is the author of several other books, including *Let's Hatch Chicks!* and *Gardening with Chickens*.

Lisa and her husband live on a small farm in Maine with a menagerie of chickens, ducks, dogs, and an indoor-outdoor barn cat. In her free time, she enjoys knitting socks, sipping herbal tea, and cooking and baking using fresh produce from the garden and fresh eggs from her coop.

Acknowledgments

It's often said that necessity is the mother of invention, and nothing could be more true when it comes to chicken keeping. I came up with much of the content of this book after trying to find a product that was safe to use around my chickens, or would do what I needed it to do, and coming up empty. I am excited to see more and more commercial products entering the market that are natural and herbal and safe. Hopefully as the years pass, more will be created, but in the meantime, these hacks will help you raise a truly happier and healthier flock. So I would like to thank my chickens for being the inspiration and reason behind much of what's contained in this book!

I would also like to thank everyone at Voyageur Press, especially my editor Thom O'Hearn, who has demonstrated superhuman patience as we worked on our third book together. Thanks also to Peg Keyser, friend and chicken photographer. From driving in snowstorms to shooting during a solar eclipse to lying on the floor of the coop waiting for a chicken to walk through the door to clicking away as we tried over and over again to get the chickens to do what we needed them to do for a shot, Peg is a consummate professional, and I couldn't be more pleased with her results!

Resources

» **Ball canning jars and silicone freezer trays—** www.freshpreserving.com

» **Beeswax pastilles—** www.bulkapothecary.com

» **Bulk grains—** www.scratchandpeck.com

» **Chalkboard paint—** www.anniesloan.com

» **Chicks and hatching eggs—** www.mypetchicken.com

» **Clark's cutting-board wax—** www.shopclarks.com

» **Coops—** www.horizonstructures.com

» **Cotton wicks—** www.bulkapothecary.com

» **Dried herbs, spices, and essential oils—** www.mountainroseherbs.com

» **Egg cartons—** www.eggcartons.com

» **Herb seedlings—** www.tastefulgarden.com

» **Herb seeds—** www.botanicalinterests.com

» **Organic garlic—** www.seedsavers.org

» **Sprouting seeds—** www.sproutpeople.org

» **Sweet-potato slips—** www.johnnyseeds.com

» **Wood glue—** www.gorillatough.com

Index